GRADUATE STUDIES IN MATHEMATICS 197

A Course on Partial Differential Equations

Walter Craig

AMS AMERICAN MATHEMATICAL SOCIETY
Providence, Rhode Island

2010 *Mathematics Subject Classification*. Primary 35-XX, 42-XX.

For additional information and updates on this book, visit
www.ams.org/bookpages/gsm-197

Library of Congress Cataloging-in-Publication Data

Names: Craig, Walter, 1953- author.
Title: A course on partial differential equations / Walter Craig.
Description: Providence, Rhode Island : American Mathematical Society, [2018] | Series: Graduate studies in mathematics ; volume 197 | Includes bibliographical references and index.
Identifiers: LCCN 2018035511 | ISBN 9781470442927 (alk. paper)
Subjects: LCSH: Differential equations, Partial–Textbooks. | AMS: Partial differential equations. msc | Harmonic analysis on Euclidean spaces. msc
Classification: LCC QA377 .C85 2018 | DDC 515/.353–dc23
LC record available at https://lccn.loc.gov/2018035511

∞ The paper used in this book is acid-free and falls within the guidelines established to ensure permanence and durability.
Visit the AMS home page at `https://www.ams.org/`

10 9 8 7 6 5 4 3 2 1 23 22 21 20 19 18

Contents

Preface

Under the Wave off Kanagawa, K. Hokusai, 1829–33.

This textbook is intended to cover material for an introductory course in partial differential equations (PDEs), aimed for fourth year undergraduate or first year graduate mathematics students. The material focuses on the three most important aspects of the subject: namely (i) theory, which is to say the questions of existence, uniqueness, and continuous dependence on given data or parameters; (ii) phenomenology, meaning the study of properties of solutions of the PDEs that we examine; and (iii) applications, by which we mean that we will make an attempt to exercise good scientific taste in choosing the topics in this text, often based on physical or geometrical applications. In taking up this subject, it is invariable that there are many editorial choices that one must make. The area of PDEs is not one unified topic; rather it is a union of numerous directions of research with different sources of problems, motivations, and goals. My own point of view in this area has been strongly influenced by my teachers and colleagues, including Louis Nirenberg, Jürgen Moser, Henry McKean, Joe Keller, and Constantine Dafermos, as well as my many friends and collaborators, including Catherine Sulem, Claude Bardos, Jean-Claude Saut, Eugene Wayne, Sergei Kuksin, and Håkan Eliasson. Members of my McMaster PDE class in 2015–2016 helped with the preparation of the text and the figures for this publication; for their input I would like to thank in particular Alexandr Chernyavsky, Nikolay Hristov, Kyle MacDonald, and Adam Sliwiak. I also want to extend a special thanks to Sarnia (Nia) Sulaiman for her rendering of numerous figures, to Ramsha Khan for her work on the images, to Emma Holmes for her proofreading, and especially to Deirdre Haskell who read the text thoroughly and critically and suggested many improvements to the presentation. However, the material itself and its presentation in this text is based on my own judgement and experiences, both scientific and pedagogical, and responsibility for omissions and incompleteness is entirely my own.

The present form of the course material follows from having given courses on PDEs many times over the years; certainly I have taught similar courses during my time on the faculty at McMaster University. But the kernel of the course material also comes from previous experiences with the graduate Mathematical Physics course at Stanford University, the two-semester sequence of courses on PDEs at Brown University, and a yearlong graduate course in PDEs at the Fields Institute in Toronto, in conjunction with the yearlong thematic program on PDEs.

Suggestions on how to use this book. The first chapters, plus a selection of the more in-depth material of the book, in particular Chapters 5–7 and sections 8.1 and 8.2, make quite a complete and interesting first semester graduate course in PDEs. If one includes the full content of Chapters 4, 6, and 8, it provides a more challenging semester-long first year graduate level course. Each chapter concludes with a set of exercises; these are on topics that augment the text, and to a large extent they are selected for their independent interest. In addition to the exercises there are a number of more challenging in-depth and open-ended projects in most chapters. Alternatively, I have taught a pleasant semester-long undergraduate course on PDEs at McMaster University based on the material in Chapters 2–4 and sections 7.1 and 7.2, including some of the material from Chapter 5 for analytic rigor. The course started with an introductory lecture that is appropriate for this level of class.

I am hopeful that the point of view adopted in this text, which brings phenomenology and applications along with the theory and intends to introduce the class to a spectrum of techniques, will be useful to students entering the field.

This work was partially supported by the Canada Research Chairs Program, the Fields Institute, and by NSERC with grant number 238452-16.

Walter Craig,
Hamilton, Ontario

Chapter 1

Introduction

Wente Torus: a genus-one surface of constant nonzero mean curvature discovered by Henry Wente in 1986. Image made in 1988 by Jim Hoffman and Yi Fang from the Geometry, Analysis, Numerics & Graphics research group at the University of Massachusetts and the Mathematical Sciences Research Institute.

1.1. Overview of the subject

It is the irony of taking university courses that one doesn't understand the real reason for studying a subject until one knows it already and has been steeped in its culture. With this paradox in mind, I will attempt to give an introduction that will motivate the material we are going to address in this course, so we can at least start with a sense of its content.

The first questions are possibly "Where do partial differential equations (PDEs) arise, and why are they useful?". In fact, the language of the sciences is mathematics (the joke has it that the language of the sciences is English with an accent). Many if not most statements in the physical sciences are in the form of mathematical equations, and the vast majority of these are differential equations, quantifying the change of one quantity in terms of others. Indeed, equations describing physical, chemical, and sometimes biological phenomena are for the most part PDEs, and the same statement holds for the engineering sciences.

Disciplines of mathematics such as geometry and dynamical systems also often give rise to PDEs. Conditions such as (i) a surface is of minimal area, (ii) a submanifold is invariant under a flow, (iii) a mapping is conformal, or (iv) the curvature tensor satisfies a particular property, are PDEs. For example, the statement that "the tensor $(g_{\mu\nu})$ is an Einstein metric for a manifold M" is a system of PDEs.

The course material will discuss the most commonly occurring PDEs and the implications they have for a function to be a solution. We are particularly interested in knowing whether solutions exist, whether they are unique, and their properties of smoothness, positivity, etc.

1.2. Examples

Here is a set of examples of some common PDEs with a brief overview of the settings where they appear, either as a characterization of mathematical properties of a geometrical object or as a description of a physical phenomenon.

Conformal mappings. A mapping between two Euclidean spaces $x \in \mathbb{R}^n \mapsto u(x) \in \mathbb{R}^N$ can be explicitly locally represented as

$$x = (x_1, \ldots, x_n) \in \mathbb{R}^n \mapsto u(x) = (u_1(x), \ldots, u_N(x)) \in \mathbb{R}^N ,$$
$$u : x \mapsto u(x) .$$

A conformal mapping is one that preserves angles. The tangent space of the image of the mapping $u(x)$ is spanned by the vectors $(\partial_{x_1} u, \ldots, \partial_{x_n} u)$. An image of a conformal mapping is given in Figure 1. Set $n = N = 2$ for our example. In the x-variables the tangent vector fields $e_{x_1} = (1, 0)$ and $e_{x_2} = (0, 1)$ are orthogonal and have the same length in the domain $\mathbb{R}^2$; a conformal mapping $u(x)$ must preserve this property for the two tangent vectors $\partial_{x_1} u$ and $\partial_{x_2} u$ in the image:

$$\partial_{x_1} u \cdot \partial_{x_2} u = 0 , \quad |\partial_{x_1} u|^2 = |\partial_{x_2} u|^2 . \tag{1.1}$$

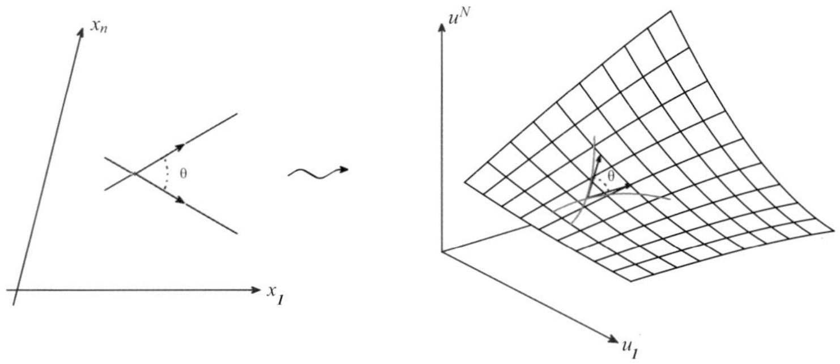

Figure 1. A conformal mapping between $\mathbb{R}^n$ and its image as a submanifold of $\mathbb{R}^N$.

Equations (1.1) are equivalent to the Cauchy–Riemann equations for either $u = (u_1(x_1, x_2), u_2(x_1, x_2))$ or $\overline{u} = (u_1(x_1, x_2), -u_2(x_1, x_2))$. That is, either u is holomorphic (analytic) in $x_1 + ix_2$,

$$\partial_{x_1} u_1 = \partial_{x_2} u_2 \ , \quad \partial_{x_2} u_1 = -\partial_{x_1} u_2 \ , \tag{1.2}$$

or else u is antiholomorphic:

$$\partial_{x_1} u_1 = -\partial_{x_2} u_2 \ , \quad \partial_{x_2} u_1 = \partial_{x_1} u_2 \ . \tag{1.3}$$

The study of such mappings is thus a central topic of complex analysis. Exercise 1.1 of this chapter is to show that (1.1) implies either (1.2) or (1.3).

Laplace's equation. Starting with the Cauchy–Riemann equations (1.2) and differentiating again, we have

$$\partial^2_{x_1} u_1 = \partial_{x_1}(\partial_{x_2} u_2) = \partial_{x_2}(\partial_{x_1} u_2) = -\partial^2_{x_2} u_1 \ ;$$

thus

$$(\partial^2_{x_1} + \partial^2_{x_2}) u_1 = 0 \ , \tag{1.4}$$

that is, u_1 is *harmonic*. The same goes for u_2 of course. The operator $\Delta = \partial^2_{x_1} + \partial^2_{x_2}$ is called the *Laplacian*; its higher-dimensional version is

$$\Delta u := (\partial^2_{x_1} + \cdots + \partial^2_{x_n}) u(x) = (\sum_{j=1}^{n} \partial^2_{x_j}) u(x) \ . \tag{1.5}$$

In general Laplace's equation is

$$\Delta u = 0 \ , \tag{1.6}$$

and solutions of (1.6) are called *harmonic* as well. This equation is a principal example of an *elliptic* equation, which will be discussed in more detail in Chapter 4. The Laplacian operator appears very often in mathematics, partially because it is the only second-order linear differential operator which is invariant under the symmetries of Euclidean space, that is, translations and rotations:

$$x \mapsto x + c \; , \qquad\qquad x \mapsto Rx \; , \tag{1.7}$$

with the rotation R represented by an orthogonal matrix $R^T = R^{-1}$. This is one elementary manifestation of the principle of relativity, which is that equations which describe physical phenomena should be invariant under symmetries of the underlying space.

The heat equation. Given coordinates of time and space as $(t, x) = (t, x_1, \dots, x_n) \in \mathbb{R}^{1+n}$, the heat equation for a function $u(t, x)$ is

$$\partial_t u = \frac{1}{2} \Delta u \; ; \tag{1.8}$$

it is the principal example of a *parabolic* equation. It was originally derived to study problems of heat conduction, and it occurs very often in probability and in problems of gradient flow, among other places.

The wave equation. Again in coordinates $(t, x) \in \mathbb{R}^{1+n}$ of space-time, the wave equation for a function $u(t, x)$ is

$$\partial_t^2 u = \Delta u \; . \tag{1.9}$$

It is a principal example of a *hyperbolic* equation; we will study a number of related equations in the next chapter, including the case of the wave equation in one space dimension.

Maxwell's equations. The propagation of electromagnetic radiation is governed by Maxwell's equations, the derivation of which is a towering achievement of nineteenth-century science. This is a system of equations coupling two vector quantities in $\mathbb{R}^3$, the electric and the magnetic vector fields:

$$\begin{aligned} E(t, x) &= (E_1(t, x), E_2(t, x), E_3(t, x)) \; , \\ B(t, x) &= (B_1(t, x), B_2(t, x), B_3(t, x)) \; . \end{aligned} \tag{1.10}$$

The six components of (E, B) satisfy the coupled system of eight equations:

$$\begin{aligned} &\partial_t E = \nabla \times B - 4\pi j \; , \\ &\partial_t B = -\nabla \times E \; , \\ &\nabla \cdot E = 4\pi\rho \; , \\ &\nabla \cdot B = 0 \; . \end{aligned} \tag{1.11}$$

The vector calculus notation is that the divergence operator $\nabla \cdot E$ is

$$\nabla \cdot E := \partial_{x_1} E_1 + \partial_{x_2} E_2 + \partial_{x_3} E_3 = \sum_{j=1}^{3} \partial_{x_j} E_j$$

and the curl is given by

$$\nabla \times B := (\partial_{x_2} B_3 - \partial_{x_3} B_2, \partial_{x_3} B_1 - \partial_{x_1} B_3, \partial_{x_1} B_2 - \partial_{x_2} B_1) \ .$$

The vector function $j(t,x) = (j_1(t,x), j_2(t,x), j_3(t,x))$ of the first line of system (1.11) represents the electric current density, and function $\rho(t,x)$ of the third line represents the electric charge density. Physical experiments have never discovered elementary particles that carry a magnetic charge analogous to electrons, protons, and other particles that carry units of electric charge. If there were such magnetic monopoles, then the other two equation components of (1.11) would have the analogous magnetic charge and current densities.

Proposition 1.1. *In the case that $\rho = 0$ and $j = 0$ (the conditions for electromagnetic wave propagation in a vacuum), Maxwell's equations* (1.11) *are equivalent to wave equations for the components of the electric and the magnetic fields.*

Proof. In the case that $\rho = 0$ and $j = 0$, one differentiates equations (1.11) to find

$$\begin{aligned} \partial_t^2 E &= \partial_t(\nabla \times B) = \nabla \times (\partial_t B) \\ &= -\nabla \times (\nabla \times E) = -(-\Delta E + \nabla(\nabla \cdot E)) \\ &= \Delta E \ . \end{aligned}$$

This is to say that each component of the electric field $E_j, j = 1,2,3$ individually satisfies the wave equation (1.9). The second line of this calculation uses a vector calculus identity and the fact that in a vacuum $\rho = 0$, and therefore the third line of Maxwell's equations is that $\nabla \cdot E = 0$. There is a similar calculation for the components of the magnetic field. □

The three equations above, namely the wave equation, the heat equation, and Laplace's equation, are the standard second-order linear constant coefficient PDEs. Most of the elementary theory of PDEs is developed to describe properties of their solutions. However, this list is far from being exhaustive; many other forms of equations, both linear and nonlinear, occur frequently and are of great interest and importance.

Monge–Ampère equations. This equation arises when the gradient of a potential function $f(x)$ is used to construct a mapping:

$$u(x) = \nabla f(x) : \mathbb{R}^n \mapsto \mathbb{R}^n \ .$$

If the mapping is constrained to be volume preserving, then $f(x)$ satisfies the Monge–Ampère equation

$$\det(\partial_{x_j} \partial_{x_\ell} f(x)) = 1 \ . \tag{1.12}$$

Similar equations play a role in the construction of Calabi–Yau metrics on manifolds. The matrix of second partial derivatives of a function $f(x)$ is known as the *Hessian* of f, and (1.12) is a fully nonlinear equation for f involving the determinant of the Hessian of f. If the function f is convex, then this is an elliptic equation. We note in comparison that the Laplacian is the trace of the Hessian

$$\Delta f = \operatorname{tr}(\partial_{x_j}\partial_{x_\ell} f) \ .$$

Schrödinger's equation. The complex valued function $\psi(t,x) = X(t,x) + iY(t,x)$ is called the *wave function* of quantum mechanics when it solves the equation

$$i\partial_t \psi = -\frac{1}{2}\Delta\psi + V(x)\psi \ . \tag{1.13}$$

This is a linear equation. The lower-order coefficient $V(x)$ is called the *potential term*; it serves to represent the environment in which the quantum particle evolves. A commonly occurring nonlinear Schrödinger equation is

$$i\partial_t \Psi = -\frac{1}{2}\Delta\Psi + c|\Psi|^2\Psi \ , \tag{1.14}$$

where the local intensity $|\Psi|^2$ of the wave function creates its own potential. This nonlinearly self-interacting potential has the effect of focusing the wave function when $c < 0$ and defocusing it when $c > 0$.

Einstein's equations. In general relativity our space-time is a manifold M^4 with a metric tensor $g = (g_{\mu\nu}(x))$ with Lorentzian signature. The Riemann curvature tensor $(R_{ijk\ell})$ is a certain nonlinear function of the metric coefficients and their derivatives, $R_{ijk\ell}(g, \partial_{x_j} g, \partial_{x_j x_\ell} g)$, as is the Ricci curvature tensor $(R_{\mu\nu})$. Einstein's equations in a vacuum consist of the system of PDEs

$$R_{\mu\nu} - \frac{1}{2}Rg_{\mu\nu} = 0 \ ,$$

where $R = \operatorname{tr} R_{\mu\nu} = \sum_{\mu\nu} R_{\mu\nu} g^{\mu\nu}$ is the scalar curvature tensor. The Einstein summation convention allows us to omit the summation symbol over repeated indices in the latter expression, so it can be represented as $R = R_{\mu\nu}g^{\mu\nu}$. When energy and/or matter are present, Einstein's equations become

$$R_{\mu\nu} - \frac{1}{2}Rg_{\mu\nu} = T_{\mu\nu} \ , \tag{1.15}$$

where T is the energy-momentum tensor.

Minimal surfaces. Given a mapping $u : D \to \mathbb{R}^N$ defined on a domain D, the surface area of the image is

$$A(u) := \int_D \sqrt{\det(\partial_{x_j} u \cdot \partial_{x_\ell} u)}\, dx \ .$$

In the case that the mapping u is given as a graph of a function $f(x)$ over $D \subseteq \mathbb{R}^n$ (where $N = n + 1$), namely that $u(x) = (x_1, \ldots, x_n, f(x))$, then

$$A(f) = \int_D \sqrt{1 + (\partial_x f)^2}\, dx \ .$$

A *minimal surface* is the result of a mapping for which the area functional $A(u)$ is minimal (or at least a critical point), generally with given additional constraints such as boundary conditions. That is to say, the variations of A about u all vanish; namely,

$$0 = \delta A(u) \cdot v := \frac{d}{d\tau}\Big|_{\tau=0} A(u + \tau v)$$

for all variations $v(x)$ such that $u + \tau v$ continues to satisfy the constraints. For the case where $n = 2$ and for a graph, if f is such a critical point, then it must satisfy the PDE

$$(1.16) \quad (1 + (\partial_{x_2} f)^2)\partial_{x_1}^2 f + (1 + (\partial_{x_1} f)^2)\partial_{x_2}^2 f - 2\partial_{x_1} f \partial_{x_2} f \partial_{x_1}\partial_{x_2} f = 0 \ .$$

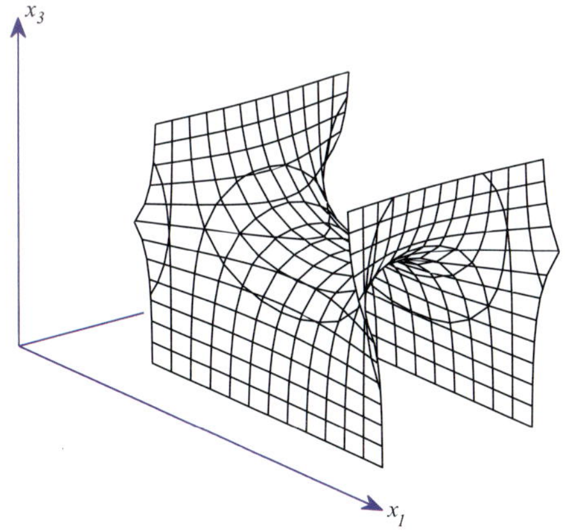

Figure 2. A minimal surface that is a graph in $\mathbb{R}^3$ whose boundary lies on a specified curve.

An example of a minimal surface is given in Figure 2, where $n = 2$ and $N = 3$. It is a certain portion of a Scherk surface, given by the equation $x_3 = \log(\frac{\cos(x_2)}{\cos(x_1)})$.

The Navier–Stokes equations. In Eulerian coordinates one describes a fluid in motion in a region $D \subseteq \mathbb{R}^n$ by its velocity vector field $u(t,x) = (u_j(t,x))_{j=1}^n$ at each point of D and at each time t. An incompressible but viscous fluid will satisfy the system of equations

$$(1.17) \qquad \begin{aligned} &\partial_t u + (u \cdot \nabla)u + \nabla p = \nu \Delta u \ , \\ &\nabla \cdot u = 0 \ . \end{aligned}$$

The *pressure* $p(t,x)$ can be thought of as the extra degrees of freedom that allow the flow determined by (1.17) to satisfy the constraint of incompressibility $\nabla \cdot u = 0$. The vorticity of the fluid is given by $\omega(t,x) := \nabla \times u$. When $n \geq 3$, it is an open question whether every solution of (1.17), even if smooth initial data is prescribed, will remain smooth at all future times.

Exercises: Chapter 1

Exercise 1.1. Prove that statement (1.1) characterizing conformal mappings implies that either (1.2) or (1.3) holds.

Exercise 1.2. Determine the class of conformal mappings in the case that $n = N \geq 3$. The conditions (1.1) are much more rigid than in two dimensions.

Exercise 1.3. Prove that the Laplacian (1.6) is invariant under Euclidean symmetries (1.7).

Exercise 1.4. In dimension $n = 2$, some elementary harmonic functions are polynomials. They include the polynomials of degrees one and two:

$$\begin{aligned} &u(x) = x_j \ , \quad j = 1, 2 \ , \\ &u(x) = x_1^2 - x_2^2 \ , \quad u(x) = x_1 x_2 \ . \end{aligned}$$

How many independent harmonic polynomials are there of general degree ℓ?

When the dimension $n \geq 3$, how many independent harmonic polynomials are there of degree ℓ?

Exercise 1.5. Prove the vector calculus identity

$$\nabla \times (\nabla \times V) = -\Delta V + \nabla(\nabla \cdot V)$$

for a vector field $V(x)$, where in our notation the gradient of a function F is

$$\nabla F = (\partial_{x_1} F, \ldots, \partial_{x_n} F) \ .$$

Chapter 2

Wave equations

Windblown ocean waves at sunrise. Image credit: Lindsay imagery / iStock / Getty Images Plus.

We will start the topic of PDEs and their solutions with a discussion of a class of wave equations, initially with several transport equations, which will lead us to the study of the standard second-order wave equation (1.9) in one space dimension. These transport equations are often called *first-order wave equations* for reasons that will become clear as we describe their solutions. We will use this study of wave equations as motivation to introduce the Fourier transform as well as to give examples of the concept of a field of characteristic curves.

2.1. Transport equations: The Fourier transform

The transport equation is a first-order equation that describes the time evolution of a quantity u, carried by a background flow field that possesses a (possibly) variable velocity $c(t,x)$, namely

$$\partial_t u + c(t,x) \cdot \partial_x u = 0 \ . \tag{2.1}$$

This typically gives rise to an *initial value problem* where one asks that

$$u(0,x) = f(x) \ .$$

The function $f(x)$ is called the *initial data*. The problem is to describe how the solution $u(t,x)$ evolves given its initial "shape" $f(x)$ at $t = 0$. Often these equations are known as first-order wave equations for the unknown function $u(t,x)$. The coefficient $c(t,x)$ is a vector known as the "speed of propagation"; we will see the reason for this when we describe the solution process.

A first example is given by the case in which the coefficient c is constant. We will see in the solution why c is known as the *speed of propagation*. We will first solve this equation using the Fourier transform. For convenience we restrict ourselves to one space dimension. The process is elementary, but nevertheless it serves to introduce a surprisingly powerful technique for the study of PDEs.

Definition 2.1. The Fourier transform of a function $f(x)$ is given by

$$\hat{f}(\xi) = \frac{1}{\sqrt{2\pi}} \int_{-\infty}^{+\infty} e^{-i\xi y} f(y)\, dy := \mathcal{F}f(\xi) \ . \tag{2.2}$$

One aspect of the power of the Fourier transform is its invertibility.

Theorem 2.2 (Fourier inversion formula). *Under suitable hypotheses on $f(x)$, there is an inversion formula for the Fourier transform*

$$f(x) = \frac{1}{\sqrt{2\pi}} \int_{-\infty}^{+\infty} e^{i\xi x} \hat{f}(\xi)\, d\xi \ .$$

At a later point we will discuss the hypotheses under which this theorem holds in much more detail; for now a perfectly reasonable condition would be that $f \in L^2(\mathbb{R})$, that is,

$$\Big(\int_{-\infty}^{+\infty} |f(x)|^2\, dx \Big)^{\frac{1}{2}} := \|f\|_{L^2(\mathbb{R})} < +\infty \ .$$

The Fourier transform is well adapted for representing derivatives.

Proposition 2.3. *Under further suitable hypotheses,*

$$\partial_x f(x) = \frac{1}{\sqrt{2\pi}} \partial_x \int_{-\infty}^{+\infty} e^{i\xi x} \hat{f}(\xi)\, d\xi = \frac{1}{\sqrt{2\pi}} \int_{-\infty}^{+\infty} \partial_x (e^{i\xi x}) \hat{f}(\xi)\, d\xi$$
$$= \frac{1}{\sqrt{2\pi}} \int_{-\infty}^{+\infty} e^{i\xi x}\, i\xi \hat{f}(\xi)\, d\xi \ .$$

That is,

$$\widehat{\partial_x f}(\xi) = i\xi \hat{f}(\xi) \ . \tag{2.3}$$

The suitable hypotheses mentioned above will be needed in order to justify the Fourier inversion formula and the exchange of the derivative ∂_x with the integral in the above calculation.

Let us assume that the propagation speed c in (2.1) is constant and that the solution $u(t,x)$ satisfies the required hypotheses of Proposition 2.3 at each time t (to ultimately have a rigorous argument, this assumption must be verified, but for the present we will not do so). Then in order to be a solution of (2.1), the function

$$u(t,x) = \frac{1}{\sqrt{2\pi}} \int_{-\infty}^{+\infty} e^{i\xi x} \hat{u}(t,\xi)\, d\xi$$

must satisfy

$$(\partial_t + c\partial_x) u(t,x) = \frac{1}{\sqrt{2\pi}} \int_{-\infty}^{+\infty} e^{i\xi x} (\partial_t \hat{u} + ci\xi \hat{u})(t,\xi)\, d\xi = 0 \ .$$

Since the Fourier transform is invertible, this means that for all $\xi \in \mathbb{R}^1$,

$$\partial_t \hat{u}(t,\xi) + ic\xi \hat{u}(t,\xi) = 0 \ ,$$

and this is a straightforward ordinary differential equation (ODE) in time for each value of the Fourier transform variable ξ:

$$\partial_t v(t) + ic\xi v(t) = 0$$

with solution

$$v(t) = e^{-ic\xi t} v(0) = e^{-ic\xi t} \hat{f}(\xi) \ .$$

Therefore an expression for our solution is

$$u(t,x) = \frac{1}{\sqrt{2\pi}} \int_{-\infty}^{+\infty} e^{i\xi x} \hat{u}(t,\xi)\, d\xi = \frac{1}{\sqrt{2\pi}} \int_{-\infty}^{+\infty} e^{i\xi x} \big(e^{-ic\xi t} \hat{f}(\xi)\big)\, d\xi$$
$$= \frac{1}{\sqrt{2\pi}} \int_{-\infty}^{+\infty} e^{i\xi(x-ct)} \hat{f}(\xi)\, d\xi = f(x-ct) \ .$$

The Fourier inversion formula is used in the last line. A picture of the evolution of a solution is given in Figure 1.

You could ask whether the identical procedure using the Fourier transform works for the case of a variable speed of propagation $c(t,x)$; it does

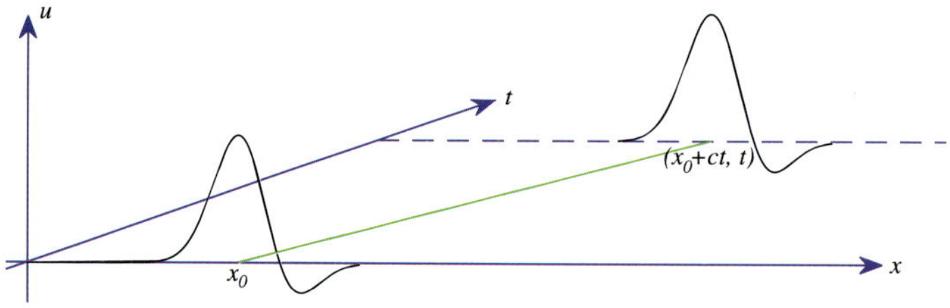

Figure 1. This is an example of the evolution of a solution of a transport equation.

not in general, nor for a time-independent speed $c(x)$. However, when $c(t)$ is independent of x, the answer is "yes" and a similar solution representation is possible. This is given as an exercise at the end of the section.

2.2. Transport equations: The method of characteristics

There is a second method for solving equations of the form (2.1) which doesn't involve the Fourier transform. Again for simplicity we will take the case that $x \in \mathbb{R}^1$. Let $f \in C^1(\mathbb{R}^1)$, the space of once continuously differentiable functions, and examine the equation

$$(\partial_t + c\partial_x)u = 0 \ , \tag{2.4}$$

again with initial data given by

$$u(0, x) = f(x) \ .$$

This is the statement that the directional derivative of $u(t, x)$ in the $(1, c)$ direction in the (t, x) plane vanishes:

$$(1, c) \cdot (\partial_t, \partial_x)u = 0 \ .$$

That is to say, $u(t, x)$ is constant along lines with tangent direction $(1, c)$. Therefore for all lines defined by x, β such that $(x - \beta) = ct$, we have $u(t, x) = u(0, \beta) = f(\beta) = f(x - ct)$. Verifying via the chain rule that this expression satisfies the transport equation (2.4) for given initial data $f \in C^1(\mathbb{R}^1)$,

$$\partial_t u(t, x) = \partial_t f(x - ct) = -cf'(x - ct) \ ,$$
$$c\partial_x u(t, x) = c\partial_x f(x - ct) = cf'(x - ct) \ ,$$

and the sum of the two terms cancels. The lines in (t, x) space defined by $(x - \beta) = ct$, exhibited in Figure 2, are the characteristic curves for this equation.

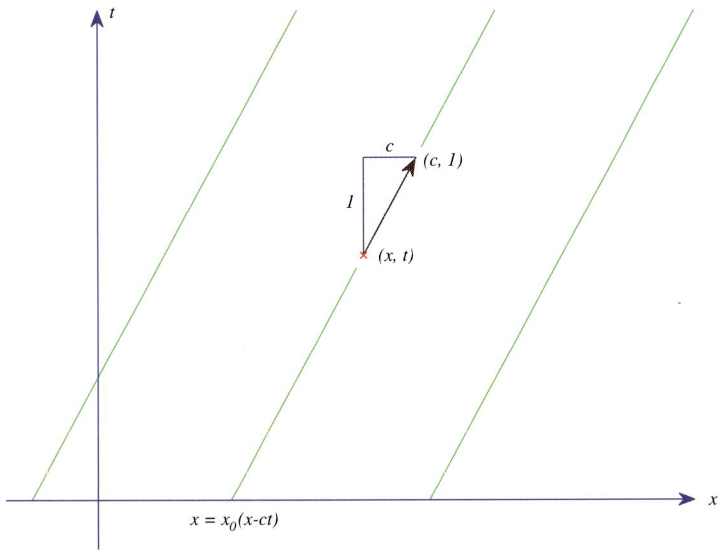

Figure 2. Geometric interpretation of a transport equation.

This observation extends to give a solution method for the general transport equation, which uses ODEs and is known as the method of characteristics. Consider the equation

$$\partial_t u + c(t,x)\partial_x u = 0 \ , \qquad u(0,x) = f(x) \ , \tag{2.5}$$

where the speed of propagation $c(t,x)$ is allowed to vary. We will assume that $c(t,x) \in C^1$. Interpret equation (2.5) in terms of directional derivatives

$$(1, c(t,x)) \cdot (\partial_t, \partial_x) u = 0 \ ,$$

which in turn is the statement that $u(t,x)$ is constant along integral curves of the vector field $X = (1, c(t,x))$ in the (t,x) plane:

$$X\, u(t,x) = (\partial_t + c(t,x)\partial_x)u = 0 \ . \tag{2.6}$$

The *characteristic equations* for the transport equation (2.5) are

$$\frac{dt}{ds} = 1 \ , \qquad \frac{dx}{ds} = c(t,x) \ , \tag{2.7}$$

with initial conditions

$$x(0) = \beta \tag{2.8}$$

whose solutions are the integral curves for X; these are the *characteristic curves* for equation (2.5). This family of curves $(t(s), x(s)) = (s, x(s,\beta))$ is parametrized by the points $\beta \in \mathbb{R}^1$ at which the curves intersect the initial surface $\{t = 0\}$. A family of characteristic curves is illustrated in Figure 3.

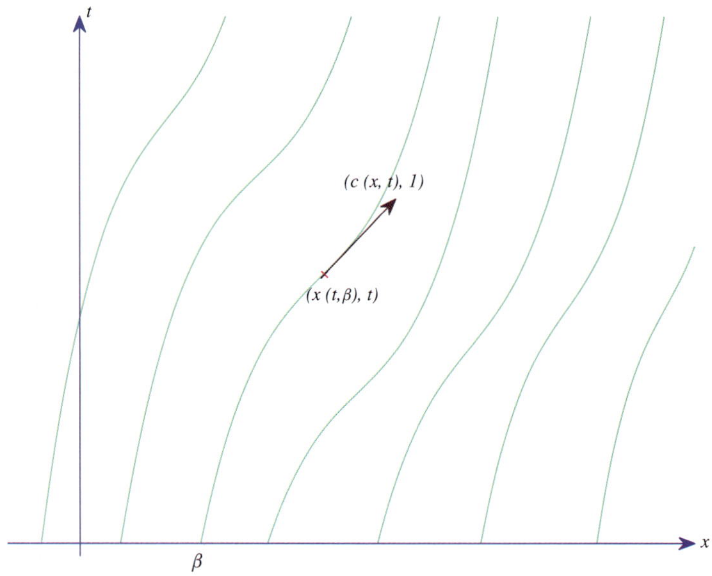

Figure 3. Characteristic curves of the vector field (2.6).

The equation states that a solution is constant along characteristic curves. Therefore, given (t, x), we define $u(t, x) = f(\beta)$, where $\beta = \beta(t, x)$ is the point of intersection with the initial surface of the characteristic curve which passes through (t, x). This involves the task of finding the point β such that the characteristic curve starting at β will pass through (t, x), basically following the characteristic curve through (t, x) backwards to the x-axis $\{t = 0\}$. To verify that our construction satisfies PDE (2.5), take into account that our proposed solution $u = u(t(s), x(s))$ is constant along a characteristic curve, and calculate using the chain rule:

$$\begin{aligned} 0 &= \frac{d}{ds} u(t(s), x(s)) \\ &= \partial_t u \frac{dt}{ds} + \partial_x u \frac{dx}{ds} = \partial_t u + c(t, x)\, \partial_x u \ . \end{aligned}$$

These equations are considered to be wave equations because they describe the propagation in time of information specified by the initial data $f(x)$ along the characteristics, at a speed given by the coefficient $c(t, x)$. Two examples of transport equations with differing global behavior are given by the following:

$$\partial_t u + x \partial_x u = 0 \ , \tag{2.9}$$

and

$$\partial_t u + x^2 \partial_x u = 0 \ . \tag{2.10}$$

Exercise 2.3 asks to solve these two equations, giving a solution representation for the initial value problem $u(0,x) = f(x)$ using explicit expressions for the respective functions $\beta(t,x)$.

2.3. Conservation laws

A *conservation law* is a first-order PDE similar to a transport equation. To be precise we are looking at linear scalar conservation laws; these take the form of an initial value problem

$$\partial_t v + \partial_x\big(c(t,x)v\big) = 0\ , \quad v(0,x) = f(x)\ . \tag{2.11}$$

If we differentiate out the second term, the equation becomes

$$\partial_t v + c(t,x)\partial_x v + \partial_x c(t,x)v = 0\ ,$$

which is a transport equation with an extra term. A related variant of a transport equation appears in Exercise 2.2. Equation (2.11) is called a conservation law because if $v(t,x) \in C^1$ is a solution of the equation, then the integral of $v(t,x)$ between any two characteristic curves $(t, x_1(t))$ and $(t, x_2(t))$ is a constant.

Proposition 2.4. *If $v(t,x) \in C^1$ solves (2.11), and $(t,x_1(t))$ and $(t,x_2(t))$ are two characteristic curves, then the quantity*

$$\int_{x_1(t)}^{x_2(t)} v(t,y)\,dy$$

is independent of t. In particular it can be determined by the initial data

$$\int_{x_1(t)}^{x_2(t)} v(t,y)\,dy = \int_{x_1(0)}^{x_2(0)} f(y)\,dy\ .$$

Proof. Recall that the characteristic curves $(t,x_1(t))$ and $(t,x_2(t))$ satisfy the characteristic equations (2.7) with initial conditions $x_1(0) = \beta_1$ and $x_2(0) = \beta_2$. The two curves do not intersect (as long as $\beta_1 \neq \beta_2$) because solutions of ODEs such as the characteristic vector field are unique. Differentiate the quantity in question with respect to time:

(2.12)

$$\partial_t\Big(\int_{x_1(t)}^{x_2(t)} v(t,y)\,dy\Big) = \int_{x_1(t)}^{x_2(t)} \partial_t v(t,y)\,dy + v(t,x_2(t))\frac{dx_2}{dt} - v(t,x_1(t))\frac{dx_1}{dt}\ .$$

Now use the fact that $x_1(t)$ and $x_2(t)$ satisfy the characteristic equations:

$$\begin{aligned} v(t,x_2(t))\frac{dx_2}{dt} - v(t,x_1(t))\frac{dx_1}{dt} &= v(t,x_2(t))c(t,x_2(t)) - v(t,x_1(t))c(t,x_1(t)) \\ &= \int_{x_1(t)}^{x_2(t)} \partial_x\big(c(t,y)v(t,y)\big)\,dy\ . \end{aligned}$$

Using this calculation in equation (2.12), we have

$$\partial_t\Big(\int_{x_1(t)}^{x_2(t)} v(t,y)\,dy\Big) = \int_{x_1(t)}^{x_2(t)} \partial_t v(t,y) + \partial_x\big(c(t,y)v(t,y)\big)\,dy = 0 \ .$$

□

Conservation laws of the form (2.11) describe the advection of a quantity $v(t,x)$ in time, such as a pollutant or a tracer, as it is carried by a background flow field given by $c(t,x)$. In this setting the function $v(t,x)$ is called the *density* of this quantity, while $c(t,x)v(t,x)$ is called the quantity's *flux*. Proposition 2.4 is the statement that the amount of quantity is conserved.

2.4. The d'Alembert formula

We now take up the study of the second-order wave equation, starting with the one space-dimensional case. In one dimension there is a particularly simple and explicit formula for solutions of the initial value problem, sometimes called the *Cauchy problem*:

$$\begin{aligned} &\partial_t^2 u - \partial_x^2 u = 0 \ , \\ &u(0,x) = f(x) \ , \quad \partial_t u(0,x) = g(x) \ . \end{aligned} \tag{2.13}$$

The initial data comprises the two functions f and g that are to be specified. Introduce new coordinates

$$x + t := r \ , \quad x - t := s \ ,$$

so that the differential operators are transformed as follows:

$$\partial_t - \partial_x = 2\partial_r \ , \quad \partial_t + \partial_x = -2\partial_s \ .$$

Then equation (2.13) is transformed to

$$(\partial_t + \partial_x)(\partial_t - \partial_x)u = -4\partial_s\partial_r u = 0 \ . \tag{2.14}$$

The meaning of equation (2.13) is now interpreted to be that $\partial_s(\partial_r u) = 0$, which is to say that $\partial_r u$ is independent of the variable s. That is,

$$\partial_r u = v(r)$$

for some function $v(r)$. Therefore, using $W = W(s)$ as an integration constant in the r-integral,

$$u(r,s) = \int^r v(r')\,dr' + W(s) := V(r) + W(s) \ .$$

Returning to the original variables (t,x), the solution takes the form

$$u(t,x) = V(x+t) + W(x-t) \ , \tag{2.15}$$

which has the interpretation of a decomposition of the solution $u(t,x)$ into a left-moving component V and a right-moving component W. That is, any solution of the one-dimensional wave equation (2.13) can be decomposed into the sum of a left-moving wave $V(x+t)$ and a right-moving wave $W(x-t)$.

The components V and W are to be expressed in terms of the initial data $u(0,x) = f(x)$ and $\partial_t u(0,x) = g(x)$. In order to do this, evaluate the expression (2.15) of the decomposition on the surface $t = 0$:

$$\begin{aligned} u(0,x) &= V(x) + W(x) = f(x) \ , \\ \partial_t u(0,x) &= V'(x) - W'(x) = g(x) \ . \end{aligned}$$

Differentiating the first line with respect to x and solving for V' and W', we find

$$V'(x) = \frac{1}{2} f'(x) + \frac{1}{2} g(x) \ , \quad W'(x) = \frac{1}{2} f'(x) - \frac{1}{2} g(x) \ ;$$

thus

$$\begin{aligned} V(x) &= \frac{1}{2} f(x) + \frac{1}{2} \int_0^x g(x')\, dx' + c_1 \ , \\ W(x) &= \frac{1}{2} f(x) - \frac{1}{2} \int_0^x g(x')\, dx' + c_2 \ , \end{aligned}$$

where necessarily $c_1 + c_2 = 0$. This is to say that

$$\begin{aligned} u(t,x) &= V(x+t) + W(x-t) \\ &= \frac{1}{2}\big(f(x+t) + f(x-t)\big) + \frac{1}{2} \int_{x-t}^{x+t} g(x')\, dx' \ . \end{aligned} \tag{2.16}$$

This is known as *d'Alembert's formula* for solutions of the one-dimensional wave equation. We note that the wave equation is a second-order PDE; however, (2.16) is well defined for data $f, g \in C(\mathbb{R}^1)$ (continuous functions) and gives a certain sense of a solution even though the resulting expression for $u(t,x)$ is not necessarily C^2. It gives a sense of a *weak solution* to the wave equation in this case, a topic that will be explored more deeply at a later point.

The solution formula (2.16) shows how the solution u at a given space-time point depends upon the initial data. Indeed, at the point (t,x) the solution $u(t,x)$ depends upon data $(f(x'), g(x'))$ only over the interval $x-t \le x' \le x+t$, and in particular it depends upon f at the two points $x \pm t$, and on $g(x')$ for $x - t \le x' \le x + t$. The set of space-time points (t,x) such that $\{x+t = x'\} \cup \{x - t = x'\}$ is called the *light cone* over x'. Given x, the region $\{x' - t = x\} \cup \{x' + t = x\}$ is called the *backwards light cone* of x. Following the d'Alembert formula, the solution $u(t,x)$ of (2.13) depends upon the initial data $(f(x), g(x))$ only for values of x' where the interior of the backwards light cone intersects the x-axis $\{t = 0\}$. This property is illustrated in Figure 4.

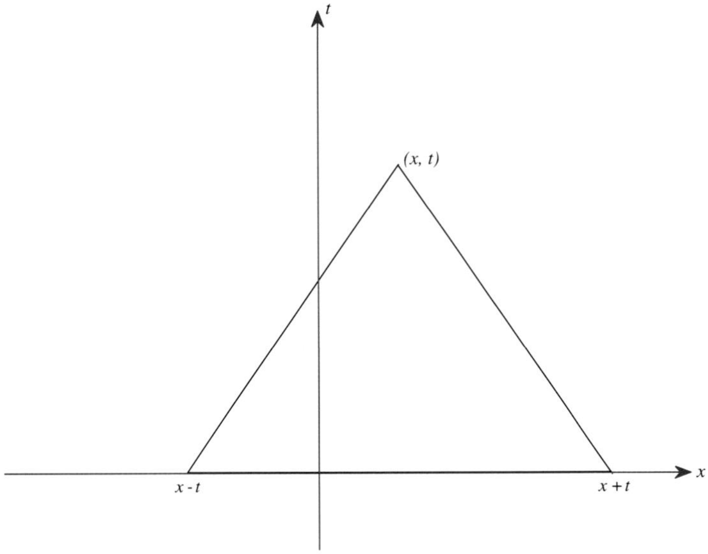

Figure 4. Solution of (2.13) using the d'Alembert formula (2.16), exhibiting the dependence on the initial data.

This is an instance of solutions satisfying the principle of *finite propagation speed.* Defining the support of a function to be

$$\mathrm{supp}(h) := \overline{\{x \in \mathbb{R}^1 \,:\, h(x) \neq 0\}}\ ,$$

and taking both (f, g) to have compact support, then if $\mathrm{supp}(f), \mathrm{supp}(g) \subseteq [-R, +R]$, then the support of the solution $u(t, x)$ satisfies $\mathrm{supp}(u(t, x)) \subseteq [-R - |t|, R + |t|]$, as can be seen in Figure 5. At time t the solution vanishes outside of the region $\{-R-|t| \leq x \leq R+|t|\}$, which is to say that information given by the initial data in the interval $[-R, +R]$ does not travel faster than speed $c = 1$ and therefore at time $t > 0$ has not propagated farther than $[-(R + t), (R + t)]$. This property of finite propagation speed is called *Huygens' principle* and is a common feature of solutions of wave equations. Furthermore, it is evident from the d'Alembert formula and Figure 5 that the solution $u(t, x)$ is constant *inside* the union of light cones emanating from the support of the initial data. Indeed, for $t > R$ and then for $|x| < t - R$,

$$u(t, x) = \frac{1}{2} \int_{x-t}^{x+t} g(x')\, dx' = \frac{1}{2} \int_{-R}^{+R} g(x')\, dx'$$

which is a constant independent of both (t, x). This property is called the *strong Huygens' principle*; it holds for solutions of the wave equation (1.9) in odd space dimensions, but not in even space dimensions nor for most other wave equations.

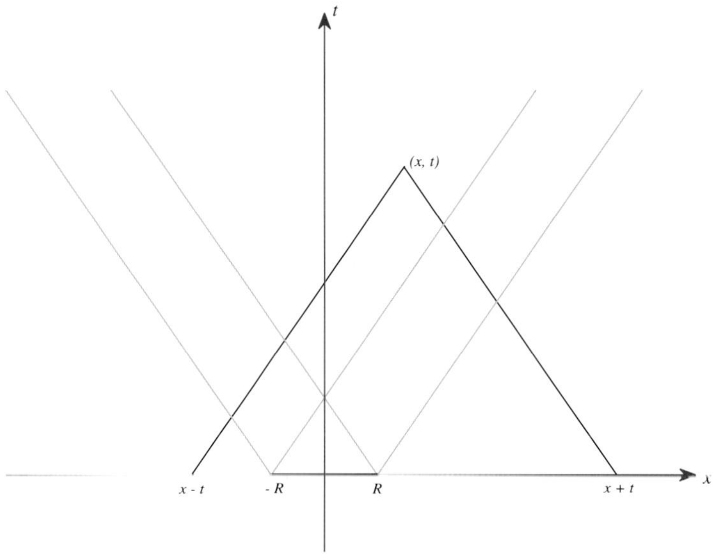

Figure 5. The d'Alembert formula (2.16) exhibits the principle of finite propagation speed and the strong Huygens' principle.

2.5. Duhamel's principle

In addition to the initial value problem for a PDE, it is also common to be given an *inhomogeneous* problem in the form

$$\partial_t^2 u - \partial_x^2 u = h(t,x) \ , \qquad (2.17)$$
$$u(0,x) = f(x) \ , \quad \partial_t u(0,x) = g(x) \ .$$

By a simple decomposition of the solution of this linear equation into two components $u(t,x) = u_1(t,x) + u_2(t,x)$, we may reduce problem (2.17) to solving for $u_1(t,x)$ given $f(x)$ and $g(x)$, with $h(t,x) = 0$, and then solving for $u_2(t,x)$ for which $f = g = 0$ and $h(t,x)$ is nonzero. A solution for $u_1(t,x)$ is given by the d'Alembert formula (2.16).

Duhamel's principle gives a method to solve for $u_2(t,x)$ that is based on a general principle. For the wave equation we first introduce the notion of a *standard form* for the initial value problem. Namely, consider the problem above,

$$\partial_t^2 u - \partial_x^2 u = 0 \ , \qquad (2.18)$$
$$u(0,x) = 0 \ , \quad \partial_t u(0,x) = g(x) \ ,$$

that is, with $f = h = 0$. Using the d'Alembert formula, the solution is given by

$$\mathbf{W}(t)(g)(x) := u(t,x) = \int_{x-t}^{x+t} g(x')\, dx' \ ,$$

which we are using to define the propagator, or solution operator $\mathbf{W}(t)$. Using this operator notation, a formula for the solution to the general initial value problem may be written

$$u(t,x) = \partial_t\big(\mathbf{W}(t)f\big)(x) + \mathbf{W}(t)g(x) \ .$$

To check that this indeed solves the wave equation with the proper initial data, one can either take the time derivative $\partial_t\big(\mathbf{W}(t)f\big)(x)$ directly on the integral expression for $\mathbf{W}(t)$, or check that for continuous $f(x)$ and $g(x)$, the desired initial values are taken on:

$$\begin{aligned}
&\partial_t\big(\mathbf{W}(0)f\big)(x) = f(x) \ , && \mathbf{W}(0)g(x) = 0 \ , \\
&\partial_t\big(\partial_t\big(\mathbf{W}(0)f\big)(x)\big) = \Delta\mathbf{W}(0)f(x) = 0 \ , && \partial_t\mathbf{W}(0)g(x) = g(x) \ .
\end{aligned}$$

We can now express a solution to the inhomogeneous problem using the propagator $\mathbf{W}(t)$. This is known as *Duhamel's principle*:

$$u_2(t,x) = \int_0^t \mathbf{W}(t-s)h(s,\cdot)(x)\,ds \ . \tag{2.19}$$

To verify that this is indeed the solution we seek, check the following:

$$\begin{aligned}
u_2(0,x) &= 0 \ , \\
\partial_t u_2(t,x) &= \partial_t \int_0^t \mathbf{W}(t-s)h(s,\cdot)(x)\,ds\Big|_{t=0} = \mathbf{W}(0)h(t,\cdot)(x) = 0 \ , \\
\partial_t^2 u_2(t,x) &= \int_0^t \partial_t^2\mathbf{W}(t-s)h(s,\cdot)(x)\,ds + \partial_t\big(\mathbf{W}(0)h(t,\cdot)(x)\big) \\
&= \Delta u_2(t,x) + h(t,x) \ .
\end{aligned}$$

In the case of the one-dimensional wave equation, using the explicit form of the operator $\mathbf{W}(t)$, the solution formula for the inhomogeneous problem (2.17) is similar to the d'Alembert formula itself. Indeed,

$$\begin{aligned}
u(t,x) &= u_1(t,x) + u_2(t,x) \\
&= \frac{1}{2}\big(f(x+t) + f(x-t)\big) + \frac{1}{2}\int_{x-t}^{x+t} g(x')\,dx' \\
&\quad + \frac{1}{2}\int_0^t \int_{x-(t-s)}^{x+(t-s)} h(s,x')\,dx'ds \ .
\end{aligned}$$

The term with the double integral accounts for the component $u_2(t,x)$ of the solution; it is the integral of $h(t,x)$ over the interior of the backwards light cone centered at the point (t,x).

Duhamel's principle stems from a formula of ODEs known as the *variation of constants.* To illustrate this, suppose we are to solve the inhomogeneous problem

$$\frac{dv}{dt} = Av + h(t) \ , \quad v(0) = g \ . \tag{2.20}$$

When $h = 0$, the solution is given by

$$v(t) = e^{tA} g \ ;$$

indeed, e^{tA} is the analog of the wave propagator $\mathbf{W}(t)$. A solution can be decomposed as follows: $v(t) = v_1(t) + v_2(t)$ where

$$v_1(t) = e^{tA} g$$

is the solution to the initial value problem with $h = 0$. The variation of constants formula gives the form of a solution to take the inhomogeneous term into account:

$$v_2(t) = \int_0^t e^{(t-s)A} h(s) \, ds \ .$$

It is straightforward to verify the following:

$$\frac{dv_2}{dt} = Av_2 + h \ , \qquad v_2(0) = 0 \ ;$$

therefore the full solution $v(t) = v_1(t) + v_2(t)$ satisfies (2.20).

2.6. The method of images

We have been discussing the initial value problem for the one-dimensional wave equation with initial data given for $x \in \mathbb{R}^1$. However, in many situations the spatial domain has boundaries on which the solution is normally required to satisfy additional conditions. Such problems are called *initial-boundary value problems.* For the one-dimensional problem, let us take the case of the spatial domain $\mathbb{R}^1_+ := \{x > 0\}$ with the boundary at $x = 0$. Two typical boundary conditions are

$$u(t, 0) = 0 \ , \tag{2.21}$$

which go by the name of *Dirichlet conditions*, and secondly

$$\partial_x u(t, 0) = 0 \ , \tag{2.22}$$

which are called *Neumann conditions.* Of course, if the support of the initial data does not include $x = 0$, by the property of finite propagation speed the solution will satisfy both of these boundary conditions for a certain time interval. However, even if so, this situation will not generally last for all time $t \in \mathbb{R}^1$.

The method of images is a classical technique used to solve the two cases of boundary conditions, (2.21) and (2.22). It also works for some other boundary conditions, which we will not pursue here. The method as applied to the problem of Dirichlet boundary conditions is based on the simple fact that a function that is odd, $h(-x) = -h(x)$, and continuous at zero must necessarily satisfy $h(0) = 0$. Since the spatial domain is $\mathbb{R}^1_+$, the initial data $f(x), g(x)$ are only defined for $x > 0$. Assume for simplicity that both $f, g \in C(\overline{\mathbb{R}^1_+})$. We require that both $f(0), g(0) = 0$ for compatibility between the initial and boundary values of the solution. The method of images consists of defining extensions $f(x) = -f(-x)$ and $g(x) = -g(-x)$ for $x < 0$, the odd reflections of $f(x)$ and $g(x)$ through the origin; these are defined and continuous on all of $x \in \mathbb{R}^1$. The d'Alembert formula gives a solution to the Cauchy problem, namely

$$u(t,x) = \frac{1}{2}\big(f(x+t) + f(x-t)\big) + \frac{1}{2}\int_{x-t}^{x+t} g(x')\,dx' \ .$$

This expression is clearly an odd function of x, as the extended values of f and g are odd. For any $x > t$, this expression only involves the arguments of f, g for positive x; however, for $x < t$, the expression involves negative arguments even though $x > 0$. To remedy this we use the fact that the extended f, g are odd to rewrite the expression as

$$u(t,x) = \frac{1}{2}\big(f(t+x) - f(t-x)\big) + \frac{1}{2}\int_{t-x}^{x+t} g(x')\,dx' \ . \tag{2.23}$$

The effect is that the solution depends only upon values of the initial data $(f(x), g(x))$ for $x > 0$. Examining this formula, it is as though the solution is reflected off of the boundary with a negative mirror image of the incident signal, as in Figure 6.

In the case of Neumann boundary conditions at $x = 0$, ask that the boundary data $f, g \in C^1(\overline{\mathbb{R}^1_+})$ and that it satisfy the compatibility conditions that

$$\partial_x f(0) = 0 = \partial_x g(0) \ .$$

Then reflect f, g to be even functions of $x \in \mathbb{R}^1$, which is possible because an even function $h(x)$ whose first derivative is continuous at $x = 0$ must satisfy $\partial_x h(0) = 0$. Using the even property in the d'Alembert formula, for $x < t$,

$$\begin{aligned} u(t,x) &= \frac{1}{2}\big(f(x+t) + f(x-t)\big) + \frac{1}{2}\int_{x-t}^{x+t} g(x')\,dx' \\ &= \frac{1}{2}\big(f(t+x) + f(t-x)\big) + \frac{1}{2}\int_{t-x}^{t+x} g(x')\,dx' + \int_0^{t-x} g(x')\,dx' \ . \end{aligned}$$

The reflection of the solution off of the boundary is now a positive mirror image of the incident signal, as illustrated in Figure 7. There is also a

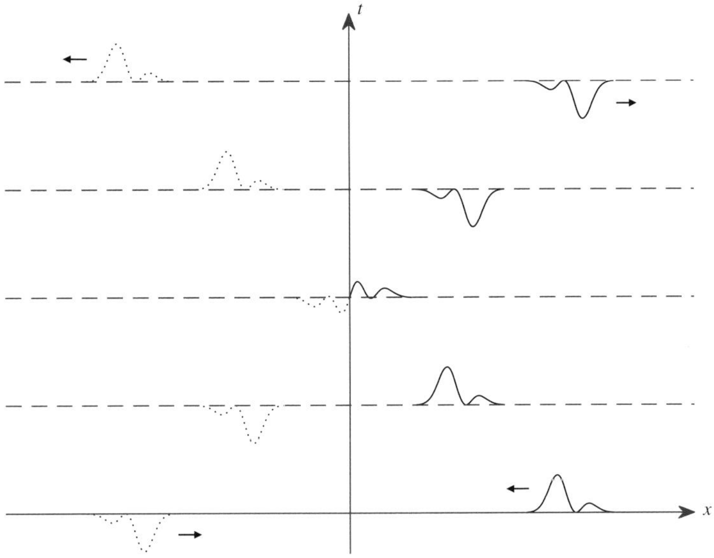

Figure 6. The method of images for Dirichlet boundary conditions.

component of the solution coming from the continuing reflection off the boundary of the contribution of g.

One further case is where nonzero boundary conditions are imposed along the t-axis; for example, one might specify the inhomogeneous data

$$u(t,0) = h(t) \tag{2.24}$$

along with the initial conditions

$$u(0,x) = f(x) , \qquad \partial_t u(0,x) = g(x) .$$

The wave equation is linear so that a sum of two solutions is also a solution of the PDEs. Therefore we may take the solution to be a sum of two components: $u_1 = u_1(t,x)$ which solves (2.13) with zero boundary data (2.21), and a second solution $u_2(t,x)$ which has homogeneous initial data $f = g = 0$ but satisfies the requested boundary conditions (2.24). The first component u_1 can be expressed as above in (2.23). The second component can be taken to vanish for $t < x$ as the backwards light cone does not intersect the boundary for positive times t. However, for $t > x$ define

$$u_2(t,x) = h(t-x) ,$$

which satisfies both the wave equation (2.13) with $f(x) = 0 = g(x)$ and the boundary conditions (2.21). The full solution is then $u(t,x) = u_1(t,x) + u_2(t,x)$. The initial and boundary data for this solution have their own compatibility conditions to satisfy. Namely, they must have $h(0) = f(0)$ to avoid discontinuities in $u(t,x)$ being introduced at $u(0,0)$, then also

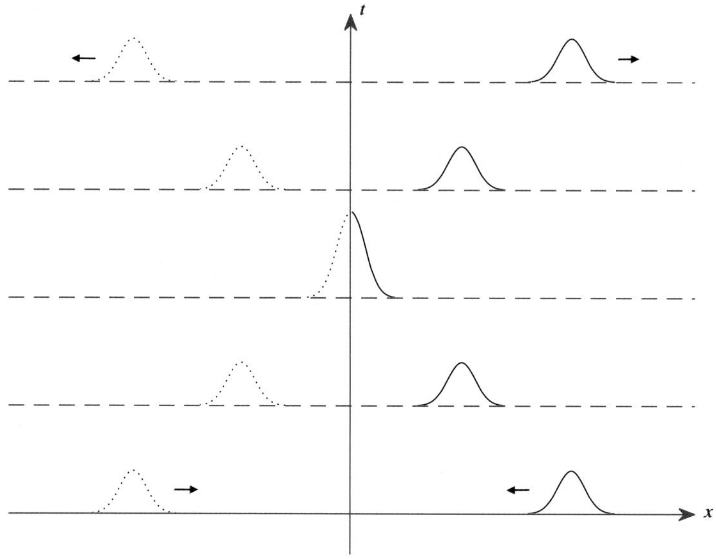

Figure 7. The method of images for Neumann boundary conditions.

$\partial_t h(0) = g(0) = \partial_t u(0,0)$, and finally $\partial_t^2 h(0) = \partial_t^2 u(0,0) = \partial_x^2 u(0,0) = \partial_x^2 f(0)$, all to avoid incompatibilities in the first and second derivatives of the solution at $(t,x) = (0,0)$.

2.7. Separation of variables

The method of separation of variables is a time-honored technique that, if one's information comes from the standard undergraduate mathematics textbook on PDEs, constitutes the entirety of the subject. In this section we will attempt to bring the topic to life by presenting it in several compelling contexts, working with the example of the wave equation. Other applications follow with straightforward modifications.

Wave equation. As a first example consider the wave equation

$$\partial_t^2 u - \partial_x^2 u = 0 \ , \quad x \in (0, 2\pi) \tag{2.25}$$

as in (2.13), with Dirichlet boundary counditions

$$u(t,0) = 0 = u(t, 2\pi) \ . \tag{2.26}$$

The premise is to seek special solutions in the form of a product $u(t,x) = v(t)w(x)$ so that

$$\partial_t^2 v \, w = v \, \partial_x^2 w \ , \tag{2.27}$$

which implies that there is a common constant λ (ignoring possible zeros of v and w) such that

$$\frac{\partial_t^2 v}{v} = \lambda = \frac{\partial_x^2 w}{w} \ .$$

Here λ is necessarily a constant because the LHS depends only upon t while the RHS depends only on x. Thus the solution form (2.27) decomposes the problem into ODEs: an evolution problem for $v(t)$ in order to satisfy the initial conditions, and an eigenvalue problem for an ordinary differential operator for $w(x)$ so that the solution satisfies the boundary conditions. The eigenvalue problem is for the pair $(w(x), \lambda)$, where the function $w(x)$ satisfies Dirichlet boundary conditions on $x \in (0, 2\pi)$:

$$\partial_x^2 w = \lambda w \ , \quad w(0) = 0 = w(2\pi) \ . \tag{2.28}$$

Given an eigenfunction and an eigenvalue from this problem, the evolution equation for v is

$$\frac{d^2}{dt^2} v = \lambda v \ ,$$

with initial data $v(0) = a$, $\dot{v}(0) = b$ to be specified. The eigenvalue problem (2.28) has a countable number of solutions, which in the case above are

$$w_k(x) = \frac{1}{\sqrt{\pi}} \sin(\frac{1}{2}kx) \ , \quad \lambda_k = -\frac{k^2}{4} \ , \quad k = 1, 2, \dots \ .$$

Thus we may choose countably many constants (a_k, b_k), $k = 1, 2, \dots$, such that

$$v_k(t) = a_k \cos(\omega_k t) + b_k \frac{\sin(\omega_k t)}{\omega_k}$$

with temporal frequencies $\omega_k = \sqrt{-\lambda_k} = \frac{1}{2}k$. For each $k \in \mathbb{N}$, this constitutes a two-parameter family of such special solutions. Linear combinations of these special solutions solve the initial value problem (2.13) with general initial data and satisfy Dirichlet boundary conditions (2.26), as the following theorem attests.

Theorem 2.5. *The general solution of the initial-boundary value problem (2.25) is a linear superposition of these special solutions:*

$$u(t, x) = \sum_{k=1}^{+\infty} \Big(a_k \cos(\omega_k t) + b_k \frac{\sin(\omega_k t)}{\omega_k}\Big) \frac{1}{\sqrt{\pi}} \sin(\frac{1}{2}kx) \ ,$$

where initial data is specified by

$$u(0, x) = f(x) = \sum_{k=1}^{+\infty} a_k \frac{1}{\sqrt{\pi}} \sin(\frac{1}{2}kx) = \sum_{k=1}^{+\infty} a_k w_k(x) \ , \tag{2.29}$$

$$\partial_t u(0, x) = g(x) = \sum_{k=1}^{+\infty} b_k \frac{1}{\sqrt{\pi}} \sin(\frac{1}{2}kx) = \sum_{k=1}^{+\infty} b_k w_k(x) \ . \tag{2.30}$$

Furthermore, the sine series used to describe $f(x), g(x)$ is complete as an orthonormal basis of the Hilbert space $L^2(0, 2\pi)$; thus any initial $f(x)$, $g(x) \in L^2$ can be represented by their Fourier sine series coefficients $(a_k, b_k)_{k\in\mathbb{N}}$, giving rise to a solution $u(t,x)$.

Given the desired initial data $f(x), g(x)$, the remaining task is to calculate the coefficients a_k of their eigenfunction expansion (2.29) for f and, respectively, b_k in the expansion (2.30) for g. It is here that the utility of an orthonormal set of basis elements becomes clear. Using our example of the sine series $w_k(x) = \frac{1}{\sqrt{\pi}}\sin(\frac{k}{2}x)$, one calculates that

$$\int_0^{2\pi} w_k(x)w_\ell(x)\,dx = \frac{1}{\pi}\int_0^{2\pi} \sin(\frac{k}{2}x)\sin(\frac{\ell}{2}x)\,dx = 0 \quad \text{for} \quad k \neq \ell\ ,$$
$$\int_0^{2\pi} w_k^2(x)\,dx = \frac{1}{\pi}\int_0^{2\pi} \sin^2(\frac{k}{2}x)\,dx = 1\ .$$

Denote the *inner product* between two functions $w(x)$ and $z(x)$ (in the Hilbert space $L^2([0,2\pi])$) by

$$\langle w, z\rangle := \int_0^{2\pi} w(x)\overline{z}(x)\,dx\ .$$

Then the expressions above for the various integrals between the eigenfunctions $w_k(x)$ are manifestations of the fact that they form an orthonormal set in $L^2([0,2\pi])$, namely $\langle w_k, w_\ell\rangle = 0$ for $k \neq \ell$ and $\langle w_k, w_k\rangle := \|w_k\|_{L^2}^2 = 1$. Theorem 2.5 also asserts that this set forms a complete set, or basis, for the Hilbert space, an important result from the theory of Fourier series (again a proof that we defer for the sake of continuity of the presentation).

To calculate the coefficients a_k, b_k of the sine series for f and g, take the inner product of (2.29) and (2.30), respectively, with the basis element w_k, and using the orthonormality relations,

$$\langle f, w_k\rangle = \langle \sum_{\ell=1}^{+\infty} a_\ell w_\ell, w_k\rangle = a_k\ ,$$
$$\langle g, w_k\rangle = \langle \sum_{\ell=1}^{+\infty} b_\ell w_\ell, w_k\rangle = b_k\ .$$

The coefficients $\{a_k\}_{k=1}^{\infty}$ are known as the Fourier sine series for the function f, and $\{b_k\}_{k=1}^{\infty}$ the Fourier sine series for g.

Quantum mechanics. There are a number of fundamental problems in mathematical physics, particularly in quantum mechanics, that have a central role in the foundations of condensed matter theory and quantum chemistry. The Schrödinger equation (1.13) for the quantum mechanical wave function $\psi(t,x)$ is one of our examples in Chapter 1. The connection between the evolution problem given in terms of the Schrödinger equation and the spectral problem presented by the Schrödinger operator is made through separation of variables. Indeed, writing

$$\psi(t,x) = e(t)\varphi(x) ,$$

the spatial dependence of the wave function is then described by a spectral problem for an ordinary or partial differential operator for the complex valued function $\varphi(x)$, along with appropriate boundary conditions. The time evolution of this special solution is given by

$$e(t) = e^{-i\lambda t}e(0)$$

for λ in the spectrum of this operator. Two settings that continue to play central roles in the physical sciences are the *quantum harmonic oscillator* and the *hydrogen atom*. The statement that the time evolution for arbitrary initial data $\psi(0,x) \in L^2$ is captured in terms of the spectral problem is a basic fact that is related to the completeness of spectral representations.

Theorem 2.6. *Let* $\mathbf{A}$ *be a self-adjoint operator on a Hilbert space* H. *Then its spectrum is contained in* $\mathbb{R}$ *and its spectral resolution is complete. If for some* $\lambda \in \mathbb{C}$ *the resolvant operator* $(\lambda\mathbf{I} - \mathbf{A})$ *is compact, then completeness means that there is a complete set of orthonormal eigenfunctions* $\{\varphi_k \in H\}$ *whose eigenvalues are real:*

$$\mathbf{A}\varphi_k = \lambda_k\varphi_k , \quad \lambda_k \in \mathbb{R} ,$$
$$\|\varphi_k\|_H^2 = 1 , \quad \langle\varphi_k, \varphi_\ell\rangle_H = 0 \quad \text{for} \quad k \neq \ell .$$

The key point to using this theorem is to show that the operator $\mathbf{A}$ is self-adjoint. For operators that are bounded on H, this is equivalent to being Hermitian symmetric:

$$\langle\varphi_1, \mathbf{A}\varphi_2\rangle_H = \langle\mathbf{A}\varphi_1, \varphi_2\rangle_H .$$

In the case of unbounded operators including differential operators such as $\mathbf{A} = \frac{d^2}{dx^2}$, self-adjointness depends upon the operator and boundary conditions. For example, for $\frac{d^2}{dx^2}$, one requires boundary conditions so that $\mathbf{A}$ is Hermitian symmetric, and one also needs to produce a dense subspace $D_A \subseteq H$ of elements of H which satisfy the boundary conditions, with the property that the graph $(D_A, \mathbf{A}(D_A)) \subseteq H \times H$ is closed. We will not give a proof of this theorem, as it lies outside the main focus of the course.

Quantum harmonic oscillator. For $x \in \mathbb{R}^1$, the quantum harmonic oscillator describes the quantum mechanical wave functions that correspond to electron states that are confined by a quadratic potential well $V(x) = \frac{1}{2}x^2$:

$$i\partial_t\psi = -\frac{1}{2}\partial_x^2\psi + \frac{1}{2}x^2\psi .$$

By the method of separation of variables, the time evolution can be reduced to be

$$\psi(t,x) = e^{-i\lambda_k t}\varphi_k(x) ,$$

where the boundary value problem for $\varphi_k(x)$ is the spectral problem

$$-\frac{1}{2}\frac{d^2}{dx^2}\varphi_k + \frac{1}{2}x^2\varphi_k = \lambda_k\varphi_k , \quad x \in \mathbb{R}^1 , \quad \psi_k \in L^2(\mathbb{R}^1) = H .$$

The eigenfunctions of this Schrödinger operator are the classical Hermite functions $\{h_k(x)\}_{k\in\mathbb{N}}$, whose eigenvalues are given by $\lambda_k = \frac{1}{2}(k+1)$, $k = 0, 1, \ldots$. The eigenfunction corresponding to the lowest eigenvalue is the ground state, in this case

$$h_0(x) = c_0 e^{-\frac{1}{2}x^2} , \quad \lambda_0 = \frac{1}{2} .$$

Hydrogen atom. In this case $x \in \mathbb{R}^3$, and one solves for the orbitals of electrons in the presence of the attractive Coulomb potential of the atomic nucleus. The Schrödinger equation describing this situation is

$$i\partial_t\psi = -\frac{1}{2}\Delta^2\psi - \frac{Z}{|x|}\psi . \tag{2.31}$$

Again, to solve this problem we separate variables in time and space to reduce the question to a spectral problem:

$$-\frac{1}{2}\Delta^2\varphi - \frac{Z}{|x|}\varphi = \lambda\varphi . \tag{2.32}$$

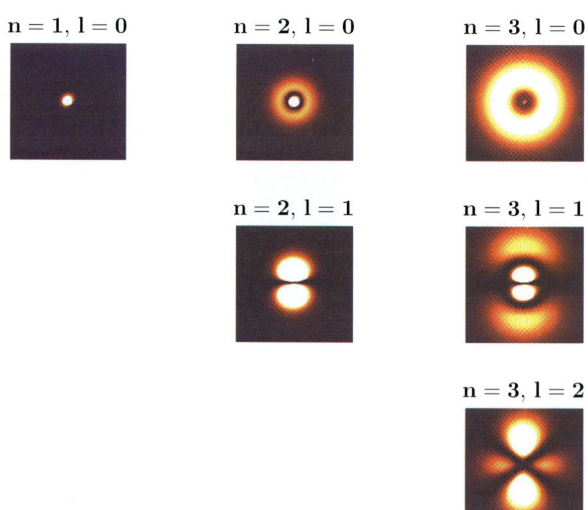

Figure 8. Cross section of probability densities $|\varphi_{n\ell m}(x)|^2$ of electron orbitals of the hydrogen atom, for the first three electronic shells.

The eigenfunctions of this Schrödinger operator are the orbitals of electrons bound to the atomic nucleus. Cross sections of the densities of the first several orbitals are given in Figure 8. The spectral lines seen in atomic spectroscopy correspond to the differences in energy between these eigenstates. Besides the obvious scientific relevance of the spectrum and the structure of the eigenstates, the interest in this spectral problem is that it also is handled through separation of the spatial variables, for which one encounters an interesting variety of classical special functions. Indeed, one writes $x = (r, \phi, \theta) \in \mathbb{R}^3$ in polar coordinates, and then making the Ansatz of a product decomposition

$$\varphi_k(x) = R_{n\ell}(r)\Phi_m(\phi)\Theta_\ell^m(\theta) ,$$

one finds that $\Phi_m(\phi)$ are given in terms of complex exponentials, while $\Theta_\ell^m(\theta)$ are associated Legendre functions of $\cos(\theta)$, and the radial dependence $R_{n\ell}(r)$ is given in terms of associated Laguerre functions. The operator (2.32) has continuous spectrum for $\lambda \geq 0$ and an infinite number of discrete negative eigenvalues $\lambda_n = -\frac{Z^2}{2n^2}$, $n = 1, 2, \ldots$, corresponding to the energies of bound electron orbitals.

Physics of music. In the hands of a good player, a musical instrument is a machine that creates a sequence of vibrations in the air that we enjoy as music. Different instruments produce these vibrations in different ways, which can be studied by modeling them mathematically.

Strings. Stringed instruments, from violins and basses to guitars to pianos, produce sounds from a bowed, plucked, or hammered string, which is a taut elastic continuum that vibrates according to the wave equation (2.25) (with units chosen so that the speed of propagation in the string is $c = 1$). The string is held fast at the bridge ($x = 0$) and at the nut or by a musician's finger on the fingerboard (at $x = 2\pi$); hence the solution satisfies Dirichlet boundary conditions. According to our analysis of the initial-boundary value problem for the wave equation above, the eigenmodes $w_k(x)$ and eigenfrequencies ω_k of the vibration of a string are given by

$$\begin{aligned} w_1(x) &= \frac{1}{\sqrt{\pi}} \sin(\frac{1}{2}x) , \quad \omega_1 = \sqrt{-\lambda_1} = \frac{1}{2} , \\ w_2(x) &= \frac{1}{\sqrt{\pi}} \sin(x) , \quad \omega_2 = \sqrt{-\lambda_2} = 1 \ \ldots \\ w_k(x) &= \frac{1}{\sqrt{\pi}} \sin(\frac{1}{2}kx) , \quad \omega_k = \sqrt{-\lambda_k} = \frac{1}{2}k . \end{aligned}$$

The first five of these eigenmodes are graphed in Figure 9.

There is certainly significance in the fact that the kth frequency satisfies $\omega_k = k\omega_1$, forming an arithmetic progression. When a string is played,

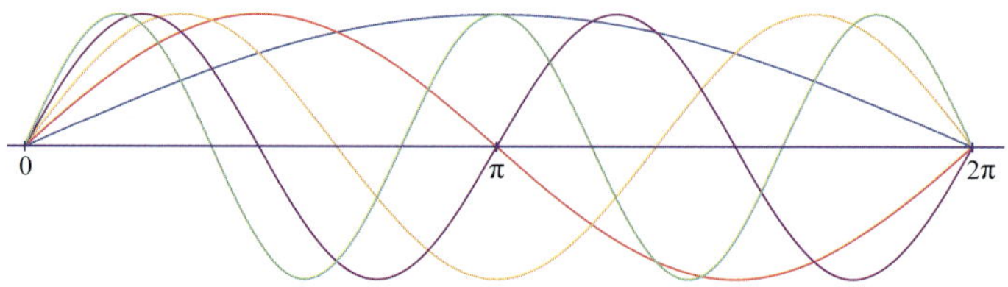

Figure 9. The first five eigenmodes of a string. Blue $= w_1(x)$, Orange $= w_2(x)$, Yellow $= w_3(x)$, Purple $= w_4(x)$, and Green $= w_5(x)$.

the resulting note is a superposition of the eigenfrequencies $\omega_1, \omega_2, \ldots$, with some weights (a_k, b_k) which generally decrease with increasing k. These are the harmonics of the fundamental tone ω_1; namely, $\omega_2 = 2\omega_1$, etc. Part of the skill of the musician is to control these weights, as our ears hear this combination of these frequencies, giving the character of the sound of the instrument. In fact, the diatonic scale of western music, as well as the pentatonic scales of eastern music and the tone selections of many other music traditions, are based very closely on these frequency relations; see Table 1.

Table 1. Overtones of a string with their correspondences to the diatonic scale.

overtone	relation to fundamental	common name
ω_1		fundamental
ω_2	$2\omega_1$	octave
ω_3	$3\omega_1$	twelfth, or major fifth plus one octave
ω_4	$4\omega_1$	second octave
ω_5	$5\omega_1$	major third above second octave
ω_6	$6\omega_1$	major fifth above second octave
ω_7	$7\omega_1$	first anharmonic overtone
ω_8	$8\omega_1$	third octave

One may check the fact that the hammers of a piano are designed to strike the string at one-seventh the distance from the end, which tends to suppress the presence of the anharmonic seventh overtone in the sound of the note.

Woodwinds. The vibrations in an air column of a wind instrument obey the Euler equations of fluid dynamics, which, in the case of the small air velocities and small pressure variations that hold in the body of the instrument, can be assumed to satisfy the linearized equation for variations around the state of constant ambient density ρ_0 and zero velocity. This is the system of equations

$$\partial_t \rho + \rho_0 \partial_x v = 0 \ , \qquad (2.33)$$
$$\rho_0 \partial_t v + p'(\rho_0)\partial_x \rho = 0$$

for small density variations $\rho(t,x)$ and small air velocities $v(t,x)$ in the air column of the instrument. The pressure $p(\rho)$ is determined as a function of ρ by a *constitutive law*, given by the theory of thermodynamics. Taking an additional time derivative of the first (or second) equation of (2.33) and then using the second (or the first, respectively), we are led to the two essentially equivalent wave equations

$$\partial_t^2 v - p'(\rho_0)\partial_x^2 v = 0 \ , \qquad (2.34)$$
$$\partial_t^2 \rho - p'(\rho_0)\partial_x^2 \rho = 0 \ ,$$

which differ through treatment at the boundaries $x = 0, 2\pi$. The wave speed is $c = \sqrt{p'(\rho_0)}$, and we will take the liberty of choosing units so that, conveniently, $c = 1$. Boundary conditions for $\rho(t,x)$, respectively for $v(t,x)$, vary according to the design of the instrument.

The flute. This instrument has large holes at the mouthpiece and at the end of the active part of the flute body, which effectively imposes that the pressure of the air column in the flute coincides with atmospheric pressure at these two boundary points; this gives boundary conditions

$$\rho(t,0) = 0 = \rho(t,2\pi) \ , \quad \text{equivalently} \quad \partial_x v(t,0) = 0 = \partial_x v(t,2\pi) \ .$$

Solving for the density $\rho(t,x)$ using the method of separation of variables gives us the eigenmodes $w_k(x) = \pi^{-1/2}\sin(\frac{1}{2}kx)$, resulting in the expression

$$\rho(t,x) = \sum_{k\geq 1} \Big(a_k \cos(\omega_k t) + b_k \frac{\sin(\omega_k t)}{\omega_k}\Big) \frac{1}{\sqrt{\pi}} \sin(\frac{1}{2}kx) \ .$$

The frequencies are clearly $\omega_k = \frac{1}{2}k$ in our chosen units, and evidently $\omega_2 = 2\omega_1$, $\omega_3 = 3\omega_1$, etc. Typically only several overtones are present in a well-played flute note (so that $a_k \sim 0$, $b_k \sim 0$ for $k > 3$ or so), reflecting the purity and simplicity of the tone of the instrument. Overblowing the flute to the second overtone to achieve the second register of the instrument increases the tone by one octave.

The clarinet. The clarinet is roughly the same length as a flute and has a similar cylindrical air column; however, the mouthpiece holds a reed in position which essentially closes this end. This imposes Dirichlet boundary conditions $v(t,0) = 0$ for the velocity, while at the bell the boundary conditions are $\partial_x v(t,2\pi) = 0$, the same as for the flute. Solving by separation of variables for the velocity in this instance, we find that the fundamental mode is $w_1(x) = \pi^{-1/2}\sin(x/4)$ with frequency $\omega_1 = 1/4$, while the higher

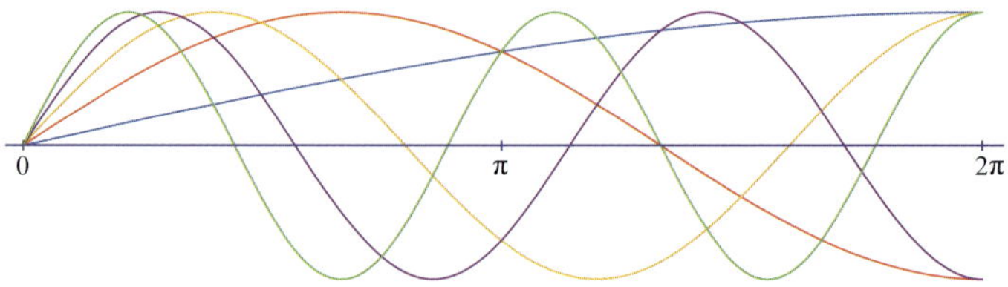

Figure 10. The first five eigenmodes of a clarinet. Blue $= w_1(x)$, Orange $= w_2(x)$, Yellow $= w_3(x)$, Purple $= w_4(x)$, and Green $= w_5(x)$.

eigenmodes are $w_k(x) = \pi^{-1/2}\sin((2k-1)x/4)$ and the frequencies of the overtones are $\omega_k = \frac{1}{2}(k-\frac{1}{2})$. The first five eigenmodes are illustrated in Figure 10. Evidently $\omega_1 = 1/4$, $\omega_2 = 3/4 = 3\omega_1$, and $\omega_3 = 5/4 = 5\omega_1$, etc. We remark that all of the even multiples of the fundamental frequency are absent and that the lowest note of a clarinet is about one octave lower than that of a flute, while the instruments are about the same size. A clarinetist overblowing to the second overtone to achieve the second register of the instrument increases the tone by one-twelfth, or one octave and a major fifth. This absence of even overtones is an explanation for the composer's rule of thumb to write the clarinet parts above the string parts. Otherwise the strings may be obliged to accord their notes with overtones that are absent in the notes of the clarinet.

The oboe and the saxophone. Both the oboe and the saxophone are reed instruments; the saxophone uses a single reed similar to the clarinet, while the oboe has a famously difficult double reed arrangement. Both impose boundary conditions at the mouthpiece and at the bell, similar to the clarinet. However, both overblow an octave rather than a twelfth. The answer to this conundrum is that they are both conical bore instruments, unlike the cylindrical bore clarinet, and must be analyzed in terms of a three-dimensional rather than one-dimensional air column. Making the reasonable assumption that the density variation in the air column (expressed in polar coordinates) is $\rho(t,r)$, namely independent of angles (θ,ϕ), the wave equation (2.34) becomes

$$\partial_t^2\rho - p'(\rho_0)(\partial_r^2 + \frac{2}{r}\partial_r)\rho = 0 \tag{2.35}$$

with boundary conditions $\partial_r\rho(t,0) = 0$ and $\rho(t,2\pi) = 0$. Making the substitution $R = r\rho(t,r)$ (d'après Beals [**3**]) leads to the equation

$$\partial_t^2 R - p'(\rho_0)\partial_r^2 R = 0 \ ,$$

satisfying boundary conditions $R(t,0) = 0 = R(t,2\pi)$. This is now addressed as before by separation of variables techniques. The fundamental is $R_1(r) = \pi^{-1/2}\sin(\frac{1}{2}r)$ with frequency $\omega_1 = \frac{1}{2}\sqrt{p'(\rho_0)}$ while, just as in the case of the

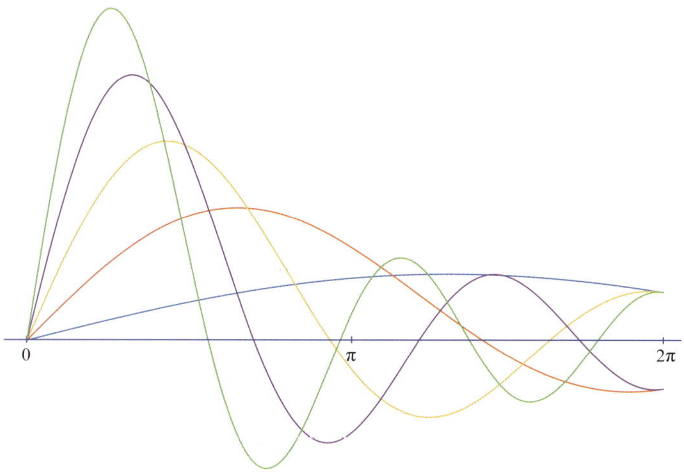

Figure 11. The first five eigenmodes of a saxophone or oboe, expressed in terms of the velocity $v(r)$ (see Project 2.14). Blue $= w_1(x)$, Orange $= w_2(x)$, Yellow $= w_3(x)$, Purple $= w_4(x)$, and Green $= w_5(x)$.

stringed instruments or the flute, the higher eigenfrequencies are $\frac{1}{2}\sqrt{p'(\rho_0)}\, k$. Again we find that $\omega_2 = 2\omega_1$, $\omega_3 = 3\omega_1$, etc. Eigenmodes in terms of the original variables are given by $\rho_n(r) = \frac{R_n(r)}{r}$. Alternatively, the problem can be phrased in terms of the velocity $v(t, r)$; see Project 2.14 at the end of this chapter. A similar analysis can be performed in terms of the velocity variables $v(r)$, giving eigenmodes $w_n(r)$ and the same sequence of eigenvalues ω_n; this is the topic of Project 2.14.

The drum. The eigenmodes of a drum correspond to those of a two-dimensional disk $B_A(0) = \{x = (x_1, x_2) : |x|^2 < A^2\}$. By separation of variables one seeks solutions $(w_k(x), \lambda_k)$ of the eigenvalue problem for Laplace's equation, which in polar coordinates is

$$(\partial_r^2 + \frac{1}{r}\partial_r)w_k + \frac{1}{r^2}\partial_\theta^2 w_k = \lambda_k w_k \ , \quad w(A, \theta) = 0 \ .$$

The angular dependence can be described in terms of complex exponentials $e^{i\ell\theta}$. However, the radial dependence involves Bessel functions whose zeros do not generally give rise to an arithmetic series of frequencies $\omega_k = \sqrt{-\lambda_k}$, as is the case for the one-dimensional instruments described above. And as we know, drums are quite noisy instruments.

Synthesizers have allowed musicians to create new instruments with the sound of a note determined by a chosen overtone series. D. DeTurck has used this idea to synthesize the sound of an imaginary piano-like instrument, where the role of the strings is replaced by higher-dimensional geometrical objects such as spheres, tori, or hyperbolic Riemann surfaces, and the subsequent overtone series of each note is determined by the square roots of the first number of Laplace–Beltrami eigenvalues. You can hear some of these instruments in action on the website `www.toroidalsnark.net/som.html`.

Exercises: Chapter 2

Exercise 2.1. For the wave speed $c(t)$ which is a general (locally integrable) function of time, give a Fourier integral representation of the solution of the transport equation (2.1).

Exercise 2.2. Find an explicit expression for the solution of the equation

$$\partial_t u + c\partial_x u + ru = 0 \ , \qquad u(0,x) = f(x) \ ,$$

using either (or both) of the above solution methods. When $r > 0$, what happens to the solution $u(t,x)$ asymptotically for $t \to +\infty$?

Exercise 2.3. Give a solution formula of the following equations using the method of characteristics:

$$\partial_t u + x\partial_x u = 0 \ , \qquad u(0,x) = f(x) \ ;$$

$$\partial_t u + x^2\partial_x u = 0 \ , \qquad u(0,x) = f(x) \ .$$

Then differentiate to check your answer. These two examples appear in the text as (2.9) and (2.10). Explain the differences between the two cases with regards to the behavior of solutions. Are the solutions $u(t,x)$ unique as functions defined over all of $(t,x) \in \mathbb{R}^2$ in both cases? What are the asymptotic states

$$v(x) = \lim_{t\to\pm\infty} u(t,x) \ ,$$

if the limit exists, that is?

Exercise 2.4. Express the solution to the following transport equation on $x \in \mathbb{R}^1$:

$$\partial_t u - xt\partial_x u = 0$$

for general initial data $u(0,x) = f(x)$.

Suppose that the initial data $f(x)$ has compact support, that is, it is zero outside of an interval $[-R, R]$. For $x \neq 0$, what is the limit

$$\lim_{t\to+\infty} u(t,x) \ ?$$

Exercise 2.5. Show that if the function $u(t,x)$ solves the transport equation (2.1), then the function $v(t,x) = \partial_x u(t,x)$ solves the conservation law (2.11).

Exercise 2.6. Give an explicit expression for the solution of the dispersive equation

$$\partial_t u \pm x\partial_x^3 u = 0 \ , \qquad u(0,x) = f(x) \ ,$$

using the Fourier transform and then the method of characteristics [4]. Comment on properties of the solution when the indicated choice of sign is $+$ and when it is $-$.

Exercise 2.7 (Conservation of energy). This problem concerns the wave equation

$$\partial_t^2 u - \partial_x^2 u = 0 \ .$$

Suppose that the initial data $f(x)$ and $g(x)$ for the wave equation satisfy

$$\int_{-\infty}^{+\infty} |\partial_x f(x)|^2 \, dx < \infty \ , \qquad \int_{-\infty}^{+\infty} |g(x)|^2 \, dx < \infty \ .$$

Define the *energy* of a solution $u(t,x)$ to be

$$E(u) = \frac{1}{2} \int_{-\infty}^{+\infty} |\partial_t u(t,x)|^2 + |\partial_x u(t,x)|^2 \, dx \ .$$

Show that for $u(t,x) \in C^2$,

$$\frac{d}{dt} E(u(\cdot,t)) = 0 \ ,$$

namely that solutions conserve the energy functional E.

Exercise 2.8 (Equipartition of energy). Consider the wave equations as in Exercise 2.7, and assume in addition that the functions $f(x)$ and $g(x)$ are compactly supported. That is, there is an interval $[-R,+R]$ such that $f(x) = 0$ and $g(x) = 0$ for all $|x| \geq R$.

Show that for $t > R$, the solution $u(t,x)$ exhibits *equipartition of energy*. Namely, for sufficiently large t,

$$\frac{1}{2} \int_{-\infty}^{+\infty} |\partial_t u(t,x)|^2 \, dx = \frac{1}{2} \int_{-\infty}^{+\infty} |\partial_x u(t,x)|^2 \, dx \ .$$

Exercise 2.9 (Traveling waves). (i) Starting from the d'Alembert formula, show that the solution of the wave equation

$$\begin{aligned} &\partial_t^2 u - \partial_x^2 u = 0 \ , \\ &u(0,x) = f(x) \ , \qquad \partial_t u(0,x) = g(x) \end{aligned}$$

can be decomposed into left-propagating and right-propagating components

$$u(t,x) = V(x+t) + W(x-t) \ .$$

(ii) Suppose that $\partial_x f(x) = -g(x)$. Show that the left-moving component of the solution vanishes.

Exercise 2.10 (Propagation of singularities). Use the d'Alembert formula to express the solution of the wave equation, $x \in \mathbb{R}^1$, with discontinuous initial data as in the following two different cases of initial data.

Case (i):

$$\begin{aligned} f(x) &= 1 \quad \text{for} \quad -1 \le x \le +1 \\ &= 0 \quad \text{for} \quad x < -1 \text{ and } 1 < x \ , \\ g(x) &= 0 \ . \end{aligned}$$

Where are the discontinuities of the solution $u(t,x)$ in space-time $(t,x) \in \mathbb{R}^2$?

Case (ii):

$$\begin{aligned} f(x) &= 0 \ , \\ g(x) &= 1 \quad \text{for} \quad -1 \le x \le +1 \\ &= 0 \quad \text{for} \quad x < -1 \text{ and } 1 < x \ . \end{aligned}$$

Is $u(t,x)$ discontinuous anywhere? Is $\partial_x u(t,x)$ discontinuous, and where?

Exercise 2.11 (Periodic boundary conditions). Use the method of separation of variables to give an expression for the wave equation

$$\partial_t^2 u - \partial_x^2 u = 0$$

on the circle $\mathbb{T}^1 := \{0 \le x < 2\pi\}$. That is, impose periodic boundary conditions on your solution in the form

$$u(t, x + 2\pi) = u(t,x) \ .$$

You may assume that $f(x), g(x)$ are 2π-periodic C^2 functions.

Exercise 2.12 (Impedence boundary conditions). Use the method of separation of variables to represent the solution of the wave equation

$$\partial_t^2 u - \partial_x^2 u = 0$$

for $0 < x < \pi$ with the following boundary conditions:

$$u(t,0) = 0 \ , \qquad \alpha \partial_x u(t,\pi) - u(t,\pi) = 0 \ .$$

The parameter α can be any real number. The behavior of a solution can be very different, depending upon α. In particular, it is different over the range $\pi < \alpha$ than over $\alpha \le \pi$. Describe this difference in terms of the stability of solutions.

Exercise 2.13 (Bloch eigenfunction expansion). Consider the Schrödinger equation (1.13) in one space dimension with $x \in \mathbb{T}^1 = \{0 \le x < 2\pi\}$, $V = 0$, and Floquet boundary conditions for the wave function

$$\psi(t, x + 2\pi) = e^{i\vartheta} \psi(t,x) \ ,$$

where ϑ is a parameter on the circle $0 \le \vartheta < 2\pi$. Use the method of separation of variables to give an expression for the solution to the Schrödinger equation in this setting.

In your analysis show that the eigenvalue problem

$$-\partial_x^2 \varphi_k(x) = \lambda_k \varphi_k(x) \ , \qquad \varphi_k(x+2\pi) = e^{i\vartheta} \varphi_k(x)$$

gives rise to a symmetric operator $\mathbf{A}$ on $L^2(\mathbb{T}^1)$ and that all of the eigenvalues $\lambda_k(\vartheta)$ are real. Indexing the eigenvalues in increasing order, show that they are 2π-periodic in ϑ. For which values of ϑ are the eigenvalues simple, and for which values are they double? Comment on the properties of continuity of these eigenvalues and of the associated eigenfunctions.

Projects: Chapter 2

Project 2.14 (Eigenmodes of the oboe). Formulated in terms of the density $\rho(t,r)$, we will assume that the air column in an oboe (or a saxophone) depends only upon the radial coordinate r, where (r, θ, ϕ) are polar coordinates in $\mathbb{R}^3$.

(i) Making the substitution $R(r) = r\rho(r)$, show that the wave equation (2.35) can indeed be rephrased for the radial function $R(r)$ as

$$\partial_t^2 R - p'(\rho_0) \partial_r^2 R = 0$$

with boundary conditions

$$R(t, 0) = 0 = R(t, 2\pi) \ .$$

See Beals [**3**] for a reference.

(ii) Using the method of separation of variables, the eigenmodes for the air column are given by an eigenvalue problem for $\rho(r)$, which in turn becomes an eigenvalue problem for $R(r)$, with evidently the eigenvalues $\lambda_n = \frac{n^2}{4}$ and eigenmodes

$$R_n(r) = \frac{1}{c_n} \sin(\frac{n}{2} r) \ .$$

Therefore $r\rho_n(r) = R_n(r)$ solves the eigenvalue problem (2.35). The L^2 inner product for radial functions defined on $B_{2\pi}(0)$ is given by

$$\langle f, g \rangle = \int_0^{2\pi} f(r) \overline{g}(r) \, r^2 dr \ .$$

Show that $\langle R_n, R_m \rangle = 0$ for $n \ne m$. Compute the value of the constant c_n such that

$$\|\rho_n\|_{L^2}^2 = \langle \rho_n, \rho_n \rangle = 1 \ .$$

(iii) The linearized Euler's equations in three dimensions can be written for (ρ, v), where ρ is the density as above and v is a velocity vector field:

$$\partial_t \rho + \rho_0 \nabla \cdot v = 0 \ ,$$
$$\rho_0 \partial_t v + p'(\rho_0) \nabla \rho = 0 \ .$$

Show that the vector field v satisfies the wave equation

$$\partial_t^2 v - p'(\rho_0) \nabla (\nabla \cdot v) = 0$$

with boundary conditions $v(t, 0) = 0$ and $\nabla \cdot v(t, 2\pi) = 0$.

(iv) The eigenvalue problem for the normal modes of the oboe can also be phrased in terms of the velocity variables v, which is a vector field in the three-dimensional problem. In polar coordinates (r, θ, ϕ) it can be written

$$v = (v_r, v_\theta, v_\phi) \ .$$

The gradient of a function f is given in polar coordinates by

$$\nabla f = \partial_r f \, \hat{r} + \frac{1}{r} \partial_\theta f \, \hat{\theta} + \frac{1}{r \sin(\theta)} \partial_\phi f \, \hat{\phi}$$

and the divergence of a vector field u is given by

$$\nabla \cdot u = \frac{1}{r^2} \partial_r (r^2 u_r) + \frac{1}{r \sin(\theta)} \partial_\theta (u_\theta \sin(\theta)) + \frac{1}{r \sin(\theta)} \partial_\phi u_\phi \ .$$

Assuming that the motion of the air column in the instrument is radial, that is,

$$v(r) = (v_r(t, r), 0, 0) \ ,$$

show that the component $v_r(t, r)$ satisfies the scalar equation

$$\partial_t^2 v_r - p'(\rho_0) \partial_r \big(\frac{1}{r^2} \partial_r (r^2 v_r) \big) = 0$$

with boundary conditions

$$v_r(t, 0) = 0 \ , \quad \frac{1}{r^2} \partial_r (r^2 v_r) = 0 \ .$$

(v) Setting $v_r = \partial_r(\frac{1}{r} W)$, show that

$$\partial_t^2 W - p'(\rho_0) \partial_r^2 W = 0$$

with boundary conditions $W(t, 0) = 0 = W(t, 2\pi)$. Thus the eigenmodes given in terms of the velocity field $w_n(r)$, corresponding to $\rho_n(r)$, are

$$w_n(r) = \frac{1}{d_n} \partial_r \big(\frac{1}{r} \sin(\frac{n}{2} r) \big) \ .$$

Show that $\langle w_n, w_m \rangle = 0$ for $n \neq m$. Calculate the normalizing constants d_n such that $\langle w_n, w_n \rangle = \|w_n\|^2 = 1$. The first five of these eigenmodes are illustrated in Figure 11.

Project 2.15 (Quantum harmonic oscillator). Solving the Schrödinger equation for the quantum harmonic oscillator

$$i\partial_t\psi = -\frac{1}{2}\partial_x^2\psi + \frac{1}{2}x^2\psi ,$$

by separation of variables one is led to the expression

$$\psi(t,x) = \sum_k a_k e^{-i\lambda_k t}\varphi_k(x) .$$

The eigenfunctions and eigenvalue pairs $(\varphi_k(x), \lambda_k)$ are given by the problem in spectral theory:

$$-\frac{1}{2}\partial_x^2\varphi_k(x) + \frac{1}{2}x^2\varphi_k(x) = \lambda_k\varphi_k(x) . \tag{2.36}$$

(i) The first eigenfunction $\psi_1(x)$ is given above as a normalized Gaussian. In the following steps, derive the orthonormalized higher eigenfunctions for (2.36), and show that the corresponding eigenvalues λ_k are all real. Verify that they are as stated in the text.

(a) For λ finite and for $|x|$ large, the potential term $V(x) = \frac{1}{2}x^2$ is dominant, and the asymptotic behavior of a solution of (2.36) is given by

$$\varphi(x) \sim ae^{x^2/2} + be^{-x^2/2} .$$

Therefore in order to have a localized solution, $a = 0$. This motivates the Ansatz that

$$\varphi(x) = e^{-x^2/2}H(x)$$

for functions $H(x)$ to be specified.

(b) Substituting this Ansatz into equation (2.36), show that

$$\frac{d^2H}{dx^2} - 2x\frac{dH}{dx} + (2\lambda - 1)H = 0 .$$

Make a formal power series expansion for $H(x)$ in the form

$$H(x) = \sum_{k=0}^{\infty} h_k x^k$$

so that

$$\frac{dH}{dx} = \sum_{k=1}^{\infty} kh_k x^{k-1} , \qquad \frac{d^2H}{dx^2} = \sum_{k=2}^{\infty} k(k-1)h_k x^{k-2} .$$

Show that the equation results in a recursion formula for the coefficients:

$$h_{k+2} = \frac{(2\lambda - 1) - 2k}{(k+1)(k+2)}h_k .$$

The first two coefficients h_0 and h_1 are not specified. Furthermore the recursion for even index coefficients h_{2m} is independent of the recursion for odd index coefficients h_{2m+1}.

(c) For most values of λ, the coefficients of this series grow rapidly and the resulting formal power series does not converge. However, the series terminates when

$$\lambda = \lambda_k = k + \frac{1}{2} \ .$$

The resulting polynomials $H(x) = H_k(x)$ are the *Hermite* polynomials, and the eigenfunctions and eigenvalues of the quantum harmonic oscillator are given by

$$\varphi_k(x) = c_k H_k(x) e^{-x^2/2} \ , \quad \lambda_k = k + \frac{1}{2} \ .$$

The normalizing constants c_k for the *Hermite functions* $h_k(x) = H_k(x)e^{-x^2/2}$ are in order to have $\|\varphi_k\|_{L^2} = 1$. In fact this gives an orthonormal basis for $L^2(\mathbb{R}^1)$. The additive constant $\frac{1}{2}$ to each eigenvalue is a correction to the original quantization postulate of Planck.

(d) Another approach to constructing the eigenfunction series $\varphi_k(x)$ is to apply a Gram–Schmidt procedure to the set of functions

$$f_\ell(x) = x^\ell e^{-x^2/2} \ .$$

One finds, to start with,

$$\varphi_0(x) = \frac{1}{\sqrt[4]{\pi}} e^{-x^2/2} \ , \quad \varphi_1(x) = \sqrt[4]{\frac{4}{\pi}} x e^{-x^2/2} \ .$$

For a continuing exercise, calculate several more normalized Hermite functions.

(ii) Show that under Fourier transform the eigenvalue problem (2.36) is transformed to itself. Conclude that the eigenfunctions of the quantum harmonic oscillator are also eigenfunctions of the Fourier transform, and vice versa.

(iii) What are the corresponding eigenvalues of the Fourier transform?

Project 2.16 (Electron orbitals for the hydrogen atom)**.** The Schrödinger equation for the motion of an electron in the quantum mechanical potential of an atomic nucleus is given in equation (2.31), which gives rise to a spectral problem for the Schrödinger operator (2.32) after a separation of variables involving time. Namely, for an eigenfunction-eigenvalue pair $(\varphi_n(x), \lambda_n)$ of the spectral problem

$$-\frac{1}{2}\Delta^2\varphi - \frac{Z}{|x|}\varphi = \lambda\varphi \ , \tag{2.37}$$

the time evolution of a solution that is initially in this eigenstate is given by

$$\psi(t, x) = e^{-i\lambda_n t}\varphi_n(x) \ .$$

Further separation of variables allows us to give a complete picture of the spectrum of the Schrödinger operator (2.32). In this process we are introduced to several of the classical special functions of mathematical physics.

(i) Because the potential function

$$V(x) = -\frac{Z}{|x|}$$

only depends upon the radial variable $|x| = r$, one transforms the problem to spherical polar coordinates $x = (r, \phi, \theta)$. In these coordinates the Laplace operator is given by

$$\Delta\varphi = \frac{1}{r^2}\partial_r\left(r^2\partial_r\varphi\right) + \frac{1}{r^2\sin^2(\theta)}\partial_\phi^2\varphi + \frac{1}{r^2\sin(\theta)}\partial_\theta\left(\sin(\theta)\partial_\theta\varphi\right) .$$

The principle of the method of separation of variables is to seek special solutions of the form

$$\varphi(r, \phi, \theta) = R(r)\Phi(\phi)\Theta(\theta) .$$

Show that a solution of (2.37) in this special form satisfies the three separate equations

$$\begin{aligned}
&\frac{d^2\Phi}{d\phi^2} = -m^2\Phi , \quad m \in \mathbb{Z} , \\
&-\frac{1}{\sin(\theta)}\frac{d}{d\theta}\left(\sin(\theta)\frac{d}{d\theta}\Theta\right) + \frac{m^2}{\sin^2(\theta)}\Theta = k\Theta , \\
&\frac{1}{2r^2}\frac{d}{dr}\left(r^2\frac{d}{dr}R\right) + \big(\lambda - V(r)\big)R = \frac{k}{2r^2}R .
\end{aligned}$$

The first of these equations evidently has the solution $\Phi(\phi) = e^{im\phi}$ for $m \in \mathbb{Z}$. Solutions $\Theta(\theta)$ for the second equation are polynomials in $\cos(\theta)$, which, after a transformation, are given by associated Legendre functions. Solutions $R(r)$ of the final equation of this list are associated Laguerre functions. These classical special functions are explored at a later point in this book, in Project 7.9 of Chapter 7.

Orbitals of electrons are the eigenfunctions of the Schrödinger operator (2.37). The physical significance of the integer m is as the z component of the angular momentum of the orbital. In the second equation, the allowed values of k are $k = \ell(\ell + 1)$ for $\ell \in \mathbb{N}$, $\ell \geq |m|$. Combining the angular behavior of the solution gives the classical spherical harmonics

$$Y_\ell^m(\phi, \theta) = \Phi_m(\phi)\Theta_\ell^m(\theta) ,$$

which are themselves the restrictions to the unit sphere in $\mathbb{R}^3$ of harmonic polynomials, a fact discussed in Exercise 1.4. The total angular momentum of the eigenstate is $L = \sqrt{\ell(\ell+1)}$. Once given $k = \ell(\ell+1)$, the third equation is a spectral problem for the function $R(r)$, which has eigenfunctions that are ordered by $n \in \mathbb{N}$ and whose eigenvalues

$$\lambda_n = -\frac{Z^2}{2n^2}$$

are determined by the solution process of the third equation. These are the energies of the eigenstates. The full orbital is given by the eigenfunction

$$\varphi(x) = R_n(r)\Phi_m(\phi)\Theta_\ell^m(\theta) \ .$$

The three numbers (n, ℓ, m) are the quantum numbers of the eigenstates; they satisfy

$$n = 1, 2, 3 \ldots \ , \quad \ell = 0, 1, 2, \ldots, n-1 \ , \quad m = -\ell, -(\ell-1) \ldots (\ell-0, \ell) \ .$$

Chapter 3

The heat equation

White hot steel on an anvil. Image credit: Fmajor / E+ / Getty Images.

The Fourier transform was originally introduced by Joseph Fourier in an 1807 paper in order to construct a solution of the heat equation on an interval $0 < x < 2\pi$, and we will also use it to do something similar for the equation

$$\begin{aligned} &\partial_t u = \frac{1}{2}\partial_x^2 u\ , \qquad t \in \mathbb{R}^1_+\ , \quad x \in \mathbb{R}^1\ , \\ &u(0,x) = f(x)\ . \end{aligned} \tag{3.1}$$

The first thing to notice is that equation (3.1) is invariant under *Brownian scaling*, which is the change of variables

$$t' = \varepsilon t\ , \quad x' = \sqrt{\varepsilon} x\ .$$

This rescaling transformation is closely related to certain principles of probability and the properties of Brownian motion. To see how the equation behaves, the change of variables gives

$$\partial_t = \varepsilon\partial_{t'} \ , \quad \partial_x = \sqrt{\varepsilon}\partial_{x'} \ ,$$

and therefore if $u(t,x)$ solves the heat equation (3.1), then so does $u' := u'(t',x') = u(\varepsilon t, \sqrt{\varepsilon}x)$:

$$\varepsilon\big(\partial_{t'} - \frac{1}{2}\partial_{x'}^2\big)u'(t',x') = \big(\partial_t - \frac{1}{2}\partial_x^2\big)u(\varepsilon t, \sqrt{\varepsilon}x) = 0 \ .$$

We will see that the ratio $x/\sqrt{t}$, which is invariant under Brownian scaling, appears in a central way in expressions for the solution, and indeed it plays a related role in probability.

3.1. The heat kernel

A derivation of the solution of (3.1) by Fourier synthesis starts with the assumption that the solution $u(t,x)$ is sufficiently well behaved so that it satisfies the hypotheses of the Fourier inversion formula. This will be verified *a posteriori* in section 5.3. Writing

$$u(t,x) = \frac{1}{\sqrt{2\pi}}\int_{-\infty}^{+\infty} e^{i\xi x}\hat{u}(t,\xi)\,d\xi \ ,$$

solutions of (3.1) must satisfy

$$0 = \big(\partial_t - \frac{1}{2}\partial_x^2\big)u(t,x) = \frac{1}{\sqrt{2\pi}}\int_{-\infty}^{+\infty} e^{i\xi x}\big(\partial_t - \frac{1}{2}(i\xi)^2\big)\hat{u}(t,\xi)\,d\xi \ .$$

We have used Proposition 2.3 to express the spatial derivatives in terms of the Fourier transform. Reasoning as before, the Fourier transform of the solution must satisfy the family of ODEs

$$\frac{d}{dt}\hat{u} + \frac{1}{2}\xi^2\hat{u} = 0 \ ,$$

parametrized by ξ. The solution is

$$\hat{u}(t,\xi) = e^{-\frac{1}{2}\xi^2 t}\hat{u}(0,\xi) = e^{-\frac{1}{2}\xi^2 t}\hat{f}(\xi) \ ,$$

where $\hat{f}(\xi)$ is the Fourier transform of the initial data. With this we can express the solution in integral operator form:

$$\begin{aligned} u(t,x) &= \frac{1}{\sqrt{2\pi}}\int_{-\infty}^{+\infty} e^{i\xi x}\big(e^{-\frac{1}{2}\xi^2 t}\hat{f}(\xi)\big)\,d\xi \\ &= \frac{1}{2\pi}\iint e^{i\xi(x-y)}e^{-\frac{1}{2}\xi^2 t}f(y)\,dyd\xi \qquad (3.2)\\ &= \int\Big(\frac{1}{2\pi}\int e^{i\xi(x-y)}e^{-\frac{1}{2}\xi^2 t}\,d\xi\Big)f(y)\,dy := \int H(t,x-y)f(y)\,dy \ . \end{aligned}$$

Justification of the exchanges of the order of integration comes from the Fubini theorem, as long as the initial data f is *absolutely integrable*, meaning that it satisfies

$$\|f\|_{L^1(\mathbb{R}^1)} := \int_{\mathbb{R}^1} |f(x)|\, dx < +\infty \ .$$

The linear space of such functions is known as $L^1(\mathbb{R}^1)$. The function $H(t, x-y)$ is the *heat kernel*, that is, the integral kernel (or the first component of the integrand) for the solution operator $\mathbf{H}(t)$ for the heat equation with the initial data $f(x)$:

$$u(t,x) = \big(\mathbf{H}(t)f\big)(x) \ .$$

Complete the square to evaluate this Fourier integral definition of the heat kernel:

$$\begin{aligned} H(t,x) &= \frac{1}{2\pi}\int_{-\infty}^{+\infty} e^{i\xi x} e^{-\frac{1}{2}\xi^2 t}\, d\xi \\ &= \frac{1}{2\pi} e^{-\frac{1}{2}x^2/t} \int_{-\infty}^{+\infty} e^{\frac{1}{2}x^2/t + i\xi x - \frac{1}{2}\xi^2 t}\, d\xi \\ &= \frac{1}{2\pi} e^{-\frac{1}{2}x^2/t} \int_{-\infty}^{+\infty} e^{-\frac{1}{2}(\xi\sqrt{t} - ix/\sqrt{t})^2}\, d\xi \ . \end{aligned}$$

Substitute $\xi' = \xi\sqrt{t} - ix/\sqrt{t}$ (which is actually an application of Cauchy's theorem of complex analysis) to find the expression

$$H(t,x) = \frac{1}{2\pi} e^{-\frac{1}{2}x^2/t} \int e^{-\frac{1}{2}(\xi')^2} \frac{d\xi'}{\sqrt{t}} = \frac{1}{\sqrt{2\pi t}} e^{-x^2/2t} \ . \tag{3.3}$$

Theorem 3.1. *A solution of the heat equation for initial data $f \in L^1(\mathbb{R}^1)$ is given by the convolution of the initial data with the heat kernel:*

$$u(t,x) = \int_{-\infty}^{+\infty} H(t, x-y) f(y)\, dy = \int_{-\infty}^{+\infty} \frac{1}{\sqrt{2\pi t}} e^{-(x-y)^2/2t} f(y)\, dy \ . \tag{3.4}$$

Take note of the invariance under Brownian scaling of the quantity

$$dP_t(x) = \frac{1}{\sqrt{2\pi t}} e^{-x^2/2t}\, dx \ ,$$

which is a Gaussian measure that is relevant to heat flow. It is a probability measure since

$$\int_{-\infty}^{+\infty} \frac{1}{\sqrt{2\pi t}} e^{-(x-y)^2/2t}\, dy = 1 \ .$$

This procedure via the heat kernel gives a unique solution among the class of bounded solutions, or among solutions which are in $L^1(\mathbb{R}^1_x)$, or simply among solutions which do not grow too rapidly as $|x| \to +\infty$. It is a surprising fact, which will come up in a later discussion, that solutions of the heat equation are not unique if we admit ones that are badly unbounded [**20**]. This topic is covered in Project 3.12 at the end of this chapter.

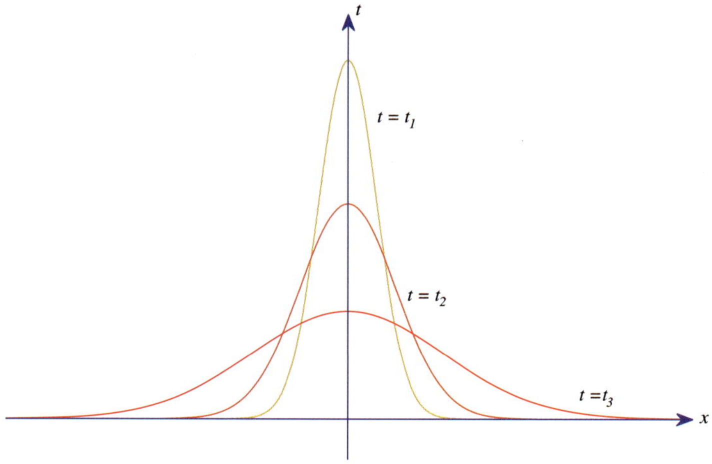

Figure 1. Superposition of the graph of the heat kernel at several times $0 < t_1 < t_2 < t_3$.

The heat kernel $H(t, x - y)$ for $t > 0$ has its maximum at $x = y$, where the maximum value is $\frac{1}{\sqrt{2\pi t}}$, and width, or standard deviation, is $\sqrt{t}$ (defined as the distance between its center and inflection point). A graph of the heat kernel taken at several snapshots in time is given in Figure 1.

It is reasonable to expect that we would obtain the same solution $u(t, x)$ at time $t > 0$ starting with data $f(x)$ at time $t = 0$ if we alternatively solved the heat equations up to time $0 < s < t$, and then, starting again at s with data $u(s, x)$, continued the solution for time $t - s$. That is, in operator notation

$$\mathbf{H}(t)f = \mathbf{H}(t - s)\mathbf{H}(s)f \ , \qquad 0 < s < t \ . \tag{3.5}$$

This is the *semigroup property*, and it is related to the question of uniqueness.

3.2. Convolution operators

The evolution operator for the heat equation is an example of a convolution operator, with convolution kernel the function $H(t, x)$.

Definition 3.2. Let $h(x) \in L^1(\mathbb{R}^1)$, and define the *convolution product* with $f(x) \in L^1(\mathbb{R}^1)$ to be

$$\big(h * f\big)(x) = \int_{-\infty}^{+\infty} h(x - y) f(y) \, dy \ .$$

In these terms the evolution operator for the heat equation can be written as

$$u(t, x) = \mathbf{H}(t) f(x) = \big(H(t, \cdot) * f\big)(x) = \int_{-\infty}^{+\infty} H(t, x - y) f(y) \, dy \ .$$

The convolution product is well defined between integrable functions:

$$\|h\|_{L^1(\mathbb{R}^1)} = \int_{\mathbb{R}^1} |h(x)|\, dx < +\infty \ .$$

In fact, whenever $f(x), h(x) \in L^1(\mathbb{R}^1)$, then $h * f(x) \in L^1(\mathbb{R}^1)$, as is shown in the following proposition.

Proposition 3.3. *Given two functions $h(x), f(x) \in L^1(\mathbb{R}^1)$, the convolution product $g(x) := h * f(x)$ is well defined. Furthermore $g(x) \in L^1(\mathbb{R}^1)$ and*

$$\|g\|_{L^1(\mathbb{R}^1)} \le \|h\|_{L^1(\mathbb{R}^1)} \|f\|_{L^1(\mathbb{R}^1)} \ .$$

Proof. These assertions will follow from the proof that $g(x)$ is integrable. To this end, estimate

$$\begin{aligned} \int_{\mathbb{R}^1} |g(x)|\, dx &= \int_{\mathbb{R}^1} \Big| \int_{\mathbb{R}^1} h(x-y) f(y)\, dy \Big|\, dx \\ &\le \int_{\mathbb{R}^1} \int_{\mathbb{R}^1} |h(x-y)| |f(y)|\, dy dx = \|h\|_{L^1(\mathbb{R}^1)} \|f\|_{L^1(\mathbb{R}^1)} \ , \end{aligned}$$

the last step following after an exchange of order of integration. □

Convolution operators with kernels $h \in L^1(\mathbb{R}^1)$ have a number of convenient algebraic features; the most elementary ones are covered in the following proposition.

Proposition 3.4. *Let $h(x), f(x), g(x) \in L^1(\mathbb{R}^1)$. Then*

(i) *The convolution product is commutative and associative:*

$$\begin{aligned} \big(h * f\big)(x) &= \big(f * h\big)(x) \ , \\ h * \big(g * f\big) &= \big(h * g\big) * f(x) \ . \end{aligned}$$

(ii) *If in addition $\partial_x f \in L^1(\mathbb{R}^1)$, then differentiation commutes with the operation of convolution:*

$$\partial_x \big(h * f\big)(x) = h * \big(\partial_x f\big)(x) \ .$$

(iii) *Under Fourier transform, convolution with $h(x)$ becomes a multiplication operator*

$$\widehat{h * f}(\xi) = \sqrt{2\pi}\hat{h}(\xi)\hat{f}(\xi) \ .$$

Proof. Address statement (iii) first:

$$\begin{aligned} &\frac{1}{\sqrt{2\pi}} \int_{-\infty}^{+\infty} e^{-i\xi x} \Big(\int_{-\infty}^{+\infty} h(x-y) f(y)\, dy \Big)\, dx \\ &= \frac{1}{\sqrt{2\pi}} \iint e^{-i\xi(x-y)} h(x-y) e^{-i\xi y} f(y)\, dy dx \end{aligned}$$

by using Fubini's Theorem. Then this quantity can be rewritten as

$$= \sqrt{2\pi}\Big(\frac{1}{\sqrt{2\pi}}\int e^{-i\xi y} f(y)\Big(\frac{1}{\sqrt{2\pi}}\int e^{-i\xi(x-y)} h(x-y)\,dx\Big)\,dy\Big)$$
$$= \sqrt{2\pi}\hat{f}(\xi)\hat{h}(\xi)\ ,$$

which is the statement of (iii). This can now be used to prove (i) and (ii). Indeed, since

$$\widehat{h * f}(\xi) = \sqrt{2\pi}\hat{h}(\xi)\hat{f}(\xi)\ ,$$

the order of multiplication does not play a role, and therefore the properties of commutativity and associativity clearly hold. Furthermore,

$$\widehat{\partial_x(h*f)}(\xi) = i\xi\widehat{h*f}(\xi) = \sqrt{2\pi}(i\xi)\hat{h}(\xi)\hat{f}(\xi)\ ,$$

and the derivative can be seen to be acting either on the convolution product or on f alone. The fact is that the Fourier transform has simultaneously diagonalized the operations of convolution with $h(x)$ and differentiation, from which the results of the proposition are corollaries. □

3.3. The maximum principle

This fundamental principle is a feature of solutions of parabolic equations such as the heat equation; we will also encounter it when we take up the topic of elliptic equations such as Laplace's equation. The following section concerns the solution $u(t,x)$ of the heat equation that is given by the heat operator $u(t,x) = (\mathbf{H}(t)f)(x) = (H(t,\cdot) * f)(x)$.

Theorem 3.5 (Maximum principle)**.** *Suppose that the initial data $f \in L^1(\mathbb{R}^1)$ satisfies $f(x) \geq 0$ for all $x \in \mathbb{R}^1$. Then either $u(t,x) > 0$ for all $t > 0$, or else both $u(t,x) \equiv 0$ and $f(x) \equiv 0$.*

The conclusion is powerful because of the fact that the existence of one zero of the solution implies that both the whole solution and the initial data must vanish identically.

Proof. Observe that the heat kernel satisfies $H(t,x) = \frac{1}{\sqrt{2\pi t}}e^{-x^2/2t} > 0$ for all $t > 0$. Hence if $f(x) \geq 0$, then clearly

$$u(t,x) = \big(H(t,\cdot) * f\big)(x) = \int_{-\infty}^{+\infty} H(t, x-y) f(y)\,dy \geq 0\ .$$

This nonstrict inequality is called the *weak maximum principle*. By a further argument we can prove that $u(t,x) > 0$ unless $f \equiv 0$. Suppose that $f(x) \geq 0$ and define sets $B_\delta = \{x : f(x) > \delta\}$. If f is not identically zero (excluding sets of zero Lebesgue measure), then B_δ is nonempty for some δ and there is a bounded set $A \subseteq B_\delta$ with positive measure. Then

$$\begin{aligned} u(t,x) &= \int_{-\infty}^{+\infty} H(t,x-y)f(y)\,dy \\ &\geq \int_A H(t,x-y)f(y)\,dy \\ &\geq \int_A H(t,x-y)\delta\,dy\ . \end{aligned}$$

This last expression is surely positive by the positive character of the heat kernel $H(t,x)$. There is more precise information available on a lower bound for $u(t,x)$ if we consider the Gaussian nature of the heat kernel:

$$u(t,x) \geq \delta\,\mathrm{meas}(A) \inf_{y\in A}\Big(\frac{1}{\sqrt{2\pi t}}e^{-\frac{(x-y)^2}{2t}}\Big)\ .$$

□

The maximum principle provides a partial ordering of solutions of the heat equation, at least for those given by the heat kernel, and thus gives a uniqueness theorem for the heat equation.

Theorem 3.6. *Suppose that $f_1(x), f_2(x) \in L^1(\mathbb{R}^1)$ and that $f_1(x) \leq f_2(x)$ for all $x \in \mathbb{R}^1$. Then the respective solutions $u_1(t,x)$, $u_2(t,x)$ either satisfy*

$$u_1(t,x) < u_2(t,x)$$

for all $t > 0$, or else $u_1(t,x) \equiv u_2(t,x)$ and $f_1(x) \equiv f_2(x)$.

Proof. The difference $u_2 - u_1$ also satisfies the heat equation, and it is given by convolution of the heat kernel with the initial data $f_2 - f_1 \geq 0$. The maximum principle asserts that either $(u_2 - u_1)(t,x) > 0$ for all $t > 0$, or else $u_2 - u_1 = 0$ and $f_2 - f_1 = 0$. □

3.4. Initial and initial-boundary value problems

The expression in (3.4) gives a solution to the heat equation, as we have derived. We now address the question as to whether this expression satisfies the desired initial conditions, that is, the conditions for which, as time t decreases to zero through positive values, the following limit holds:

$$\lim_{t\to 0^+} u(t,x) = \lim_{t\to 0^+} \int_{-\infty}^{+\infty} H(t,x-y)f(y)\,dy = f(x)\ .$$

There is a satisfactory answer at points $x_0 \in \mathbb{R}^1$ at which $f(x)$ is continuous. Exercise 3.5 addresses the case in which the initial data has a jump discontinuity at $x = x_0$.

Theorem 3.7. *Suppose that $f \in L^1(\mathbb{R}^1)$ and additionally that $f(x)$ is continuous at a point $x = x_0$. Then*

$$\lim_{t \to 0+} \int_{-\infty}^{+\infty} H(t, x_0 - y) f(y)\, dy = f(x_0) \ .$$

This is an instance of a Dirac δ-function. The δ-function is expressed formally as a distribution that satisfies the relation

$$\int_{-\infty}^{+\infty} \delta(x_0 - y) f(y)\, dx = f(x_0)$$

for any continuous function $f(x)$. Its mathematical sense is defined in terms of a limit, as in the limit of the statement of the theorem.

Proof. Since we know $f(x)$ to be continuous at $x = x_0$, for any $\varepsilon > 0$ there is a $\delta > 0$ such that

$$|f(x) - f(x_0)| < \frac{\varepsilon}{3} \tag{3.6}$$

for all $|x - x_0| < \delta$. Secondly, we recall the fact that

$$\int_{-\infty}^{+\infty} \frac{1}{\sqrt{2\pi t}} e^{-\frac{(x_0 - y)^2}{2t}}\, dy = 1$$

for all $t > 0$ because of the normalization of the heat kernel. Furthermore, for $y \neq x_0$ the heat kernel satisfies

$$\lim_{t \to 0+} \frac{1}{\sqrt{2\pi t}} e^{-\frac{(x_0 - y)^2}{2t}} = 0 \ .$$

This limit is uniform for y bounded away from x_0, and in fact for $\delta > 0$ fixed, for any $\varepsilon > 0$ there exists a time $t(\varepsilon) > 0$ such that for any smaller $0 < t < t(\varepsilon)$, both

$$\frac{1}{\sqrt{2\pi t}} e^{-\frac{(x_0 - y)^2}{2t}} \|f\|_{L^1} < \frac{\varepsilon}{3} \quad \text{for any} \quad |y - x_0| \geq \delta \ , \tag{3.7}$$

$$\int_{|y - x_0| \geq \delta} \frac{1}{\sqrt{2\pi t}} e^{-\frac{(x_0 - y)^2}{2t}}\, dy\, |f(x_0)| < \frac{\varepsilon}{3} \ . \tag{3.8}$$

Now split up the integral in expression (3.4) for the solution of the heat equation in the following way. Given $\varepsilon > 0$, choose $\delta > 0$ so that (3.6) holds for $|x - x_0| < \delta$, and consider

$$
\begin{aligned}
f(x_0) &- \int_{-\infty}^{+\infty} \frac{1}{\sqrt{2\pi t}} e^{-\frac{(x_0-y)^2}{2t}} f(y)\, dy \\
&= \int_{-\infty}^{+\infty} \frac{1}{\sqrt{2\pi t}} e^{-\frac{(x_0-y)^2}{2t}} [f(x_0) - f(y)]\, dy \\
&= \int_{-\infty}^{x_0-\delta} + \int_{x_0+\delta}^{+\infty} \frac{1}{\sqrt{2\pi t}} e^{-\frac{(x_0-y)^2}{2t}} [f(x_0) - f(y)]\, dy \\
&\quad + \int_{x_0-\delta}^{x_0+\delta} \frac{1}{\sqrt{2\pi t}} e^{-\frac{(x_0-y)^2}{2t}} [f(x_0) - f(y)]\, dy\ .
\end{aligned}
$$

For $0 < t < t(\varepsilon)$, the first two integrals are bounded in terms of ε using, respectively, (3.8) for the terms containing $f(x_0)$ and (3.7) for those containing $f(y)$:

$$
\begin{aligned}
&\Big| \int_{-\infty}^{x_0-\delta} + \int_{x_0+\delta}^{+\infty} \frac{1}{\sqrt{2\pi t}} e^{-\frac{(x_0-y)^2}{2t}} [f(x_0) - f(y)]\, dy \Big| \\
&\le |f(x_0)| \Big| \int_{-\infty}^{x_0-\delta} + \int_{x_0+\delta}^{+\infty} \frac{1}{\sqrt{2\pi t}} e^{-\frac{(x_0-y)^2}{2t}}\, dy \\
&\quad + \int_{-\infty}^{+\infty} |f(y)|\, dy \sup_{|x_0-y|\ge\delta} \frac{1}{\sqrt{2\pi t}} e^{-\frac{(x_0-y)^2}{2t}} \\
&\le \frac{2\varepsilon}{3}\ .
\end{aligned}
$$

Using (3.6), the remaining integral over the interval $[x_0 - \delta, x_0 + \delta]$ is also shown to be bounded in terms of ε:

$$
\begin{aligned}
&\Big| \int_{x_0-\delta}^{x_0+\delta} \frac{1}{\sqrt{2\pi t}} e^{-\frac{(x_0-y)^2}{2t}} [f(x_0) - f(y)]\, dy \Big| \le \frac{\varepsilon}{3} \int_{x_0-\delta}^{x_0+\delta} \frac{1}{\sqrt{2\pi t}} e^{-\frac{(x_0-y)^2}{2t}}\, dy \\
&\le \frac{\varepsilon}{3} \int_{-\infty}^{+\infty} \frac{1}{\sqrt{2\pi t}} e^{-\frac{(x_0-y)^2}{2t}}\, dy = \frac{\varepsilon}{3}\ .
\end{aligned}
$$

Since ε can be chosen arbitrarily small, this proves the statement of the theorem. □

Method of images. As with the wave equation, there are methods for solving the heat equation in a region with a boundary, where boundary conditions are imposed. Not surprisingly, this involves the method of images. The two basic boundary conditions are homogeneous Dirichlet conditions, which we will impose at $x = 0$ for the region $\{x > 0\} := \mathbb{R}^1_+$,

$$\begin{aligned} &\partial_t u = \frac{1}{2}\partial_x^2 u \ , \quad x > 0 \ , t > 0 \ , \\ &u(0, x) = f(x) \ , \quad x > 0 \ , \\ &u(t, 0) = 0 \ , \end{aligned}$$

and homogeneous Neumann boundary conditions

$$\partial_x u(t, 0) = 0 \ .$$

In physical terms, homogeneous Dirichlet conditions correspond to the boundary being held at a fixed temperature, such as the reference temperature $u(t, 0) = 0$, as though it were connected to an infinite reservoir of heat energy at this temperature. Homogeneous Neumann boundary conditions correspond to the boundary being a perfect heat insulator, allowing no heat flux through it.

Initial data $f(x)$ is given for $x > 0$; the method of images corresponds to extending the initial data to $x < 0$ in a way that the resulting solution will obey the desired boundary conditions. In the case of Dirichlet boundary conditions we reflect the data to be odd in x; namely, for $x < 0$ define $f(x) = -f(-x)$. Then the solution of the heat equation on the whole line is given by

$$\begin{aligned} u(t, x) &= \int_{\mathbb{R}^1} H(t, x - y) f(y)\, dy \\ &= \int_{y \geq 0} H(t, x - y) f(y)\, dy + \int_{y < 0} H(t, x - y) f(y)\, dy \\ &= \int_{y \geq 0} [H(t, x - y) - H(t, x + y)] f(y)\, dy \ . \end{aligned}$$

The heat kernel for the Dirichlet problem on $\mathbb{R}^1_+$ is therefore

$$H_D(t, x, y) = [H(t, x - y) - H(t, x + y)] = \frac{1}{\sqrt{2\pi t}}[e^{\frac{(x-y)^2}{2t}} - e^{\frac{(x+y)^2}{2t}}] \ ,$$

where both $x > 0$ and $y > 0$. We may denote the solution operator for heat flow on R^1_+ with Dirichlet boundary conditions by

$$\mathbf{H}_D(t) f(x) = \int_0^{+\infty} H_D(t, x, y) f(y)\, dy \ .$$

Similarly, initial data for the case of Neumann boundary conditions use the method of images with an even reflection of the initial data, defining

$f(x) = f(-x)$ for $x < 0$. In this case

$$\begin{aligned} u(t,x) &= \int_{\mathbb{R}^1} H(t, x-y) f(y)\, dy \\ &= \int_{y \geq 0} H(t, x-y) f(y)\, dy + \int_{y<0} H(t, x-y) f(y)\, dy \\ &= \int_{y \geq 0} [H(t, x-y) + H(t, x+y)] f(y)\, dy \ . \end{aligned}$$

The resulting heat kernel for the Neumann problem on $\mathbb{R}^1_+$ is

$$H_N(t,x,y) = [H(t, x-y) + H(t, x+y)] = \frac{1}{\sqrt{2\pi t}}[e^{\frac{(x-y)^2}{2t}} + e^{\frac{(x+y)^2}{2t}}] \ ,$$

where again both $x > 0$ and $y > 0$. One may similarly denote the solution operator for heat flow on R^1_+ with Neumann boundary conditions by

$$\mathbf{H}_N(t) f(x) = \int_0^{+\infty} H_N(t,x,y) f(y)\, dy \ .$$

Duhamel's principle. The inhomogeneous problem for the heat equation is the following:

$$\begin{aligned} \partial_t u - \frac{1}{2}\partial_x^2 u &= h(t,x) \ , \\ u(0,x) &= f(x) \ , \end{aligned} \tag{3.9}$$

which we consider first for $x \in \mathbb{R}^1$. The general principle given in section 2.5 indicates the solution process. As before, we decompose the solution as $u(t,x) = u_1(t,x) + u_2(t,x)$ where u_1 is a solution of the initial value problem with initial data $f(x)$, for which we take the inhomogeneous term to be zero. The component $u_2(t,x)$ then accounts for the presence of a nonzero $h(t,x)$, which, following Duhamel's principle, is given by the expression

$$\begin{aligned} u_2(t,x) &= \int_0^t \mathbf{H}(t-s) h(s,\cdot)(x)\, ds \\ &= \int_0^t \int_{-\infty}^{+\infty} \frac{1}{\sqrt{2\pi(t-s)}} e^{\frac{(x-y)^2}{2(t-s)}}\ h(s,y)\, dyds \ . \end{aligned}$$

One can check that $u_2(0,x) = 0$, so that the component $u_1(t,x)$ accounts for the initial data $f(x)$, and that

$$\partial_t u_2 = \frac{1}{2}\partial_x^2 u_2 + h(t,x) \ .$$

The inhomogeneous problems on $\mathbb{R}^1_+$ have similar solution formulae. This problem with Dirichlet boundary conditions imposed at $x = 0$ is

$$\partial_t u = \frac{1}{2}\partial_x^2 u + h(t,x) , \qquad x > 0 ,$$
$$u(0,x) = f(x) , \quad x > 0 , \qquad u(t,0) = 0 .$$

Decomposing the solution $u(t,x) = u_1(t,x) + u_2(t,x)$ as above, and using the solution operator $u_1 = \mathbf{H}_D(t)f$ for the first component, the second component is given by Duhamel's principle:

$$\begin{aligned} u_2(t,x) &= \int_0^t \mathbf{H}_D(t-s)h(s,\cdot)\,ds \\ &= \int_0^t H_D(t-s,x,y)h(s,y)\,dyds . \end{aligned}$$

Similarly, one may instead impose Neumann boundary conditions at the boundary $x = 0$:

$$\partial_x u(t,0) = 0 .$$

In this case the Neumann heat kernel is used so that $u_1(t,x) = \mathbf{H}_N(t)f$ and

$$u_2(t,x) = \int_0^t H_N(t-s,x,y)h(s,y)\,dyds .$$

Green's identities. The heat kernel $H(t-s, x-y)$ that appears in the formulae of Duhamel's principle, considered as a function of (t,x), is a solution of the heat equation. Considered as a function of (s,y), it is a solution of the heat equation with reversed time $0 < s < t$:

$$\partial_s H = -\frac{1}{2}\partial_y^2 H . \tag{3.10}$$

Given any functions $G(s,y)$ and $u(s,y)$ that are both smooth in the domain $x \in \mathbb{R}^1$ and $0 < s < t$, integrations by parts result in the following calculus identities:

$$\begin{aligned} &\int_0^t \int_{-\infty}^{+\infty} G(s,y)\big(\partial_s u - \frac{1}{2}\partial_y^2 u\big)\,dyds \\ &= \int_0^t \int_{-\infty}^{+\infty} (-\partial_s G - \frac{1}{2}\partial_y^2 G)u(s,y)\,dyds + \int_{-\infty}^{+\infty} G(s,y)u(s,y)\,dy\Big|_{s=0}^{t} , \end{aligned}$$

where we will assume that G and u decrease to zero as $|y| \to +\infty$ sufficiently rapidly so that there are no remaining boundary terms. If $G(s,y)$ also satisfies (3.10), then the RHS of the integral expression simplifies. If we take $G(s,y) = H(t-s, x-y)$, we also know from Theorem 3.7 that

$$\lim_{s\to t^-} \int_{-\infty}^{+\infty} H(t-s,x-y)u(s,y)\,dy = u(t,x) ,$$

and therefore

$$\int_0^t \int_{-\infty}^{+\infty} H(t-s, x-y)\big(\partial_s u - \frac{1}{2}\partial_y^2 u\big)\, dyds \\ = u(t,x) - \int_{-\infty}^{+\infty} H(t, x-y) u(0,y)\, dy \ . \tag{3.11}$$

This is known as *Green's identity* for the heat equation. From it we recover the statement that if $u(t,x)$ is a solution of the heat equation (3.1) which decays at $|x| \to +\infty$ sufficiently well, then

$$u(t,x) = \int_{\infty}^{+\infty} H(t, x-y) f(y)\, dy \ .$$

Furthermore, if $u(t,x)$ is a solution of the inhomogeneous heat equation (3.9), then this gives an alternate derivation of the formula stemming from Duhamel's principle,

$$u(t,x) = \int_{-\infty}^{+\infty} H(t, x-y) f(y)\, dy + \int_0^t \int_{-\infty}^{+\infty} H(t-s, x-y) h(s,y)\, dyds \ .$$

Inhomogeneous boundary conditions. Green's identities are also useful for the initial-boundary value problem for the heat equation. The analog of the calculus identities for the initial-boundary value problem with given Dirichlet boundary data is that

$$\begin{aligned}
&\int_0^t \int_0^{+\infty} H_D(t,x,y)\big(\partial_s u - \frac{1}{2}\partial_y^2 u\big)\, dyds \\
&\quad = \int_0^t \int_0^{+\infty} \big(-\partial_s H_D - \frac{1}{2}\partial_y^2 H_D\big) u(s,y)\, dyds \\
&\quad + \int_0^{+\infty} H_D(t-s, x, y) u(s,y)\, dy\Big|_{s=0}^{t} \\
&\quad + \frac{1}{2}\int_0^t \Big(-H_D(t-s, x-y)\partial_y u - \partial_y H_D(t-s, x-y) u\Big)\, ds\Big|_{y=0}^{+\infty} \ .
\end{aligned}$$

The heat kernel $H_D(t-s, x, y)$ satisfies (3.10). We also know that the limit as $s \to t_-$ of the heat kernel $H_D(t, s, x, y) = \delta_x(y)$. Using this as well as the fact that $H_D(t-s, x, 0) = 0$, and furthermore assuming that the solution $u(s,y)$ does not grow faster than the decay of the heat kernel as $y \to +\infty$, the above formula gives us a second version of Green's identities (3.11), namely

$$\begin{aligned}
&\int_0^t \int_0^{+\infty} H_D(t,x,y)\big(\partial_s u - \frac{1}{2}\partial_y^2 u\big)\, dyds \\
&\qquad = u(t,x) - \int_0^{+\infty} H_D(t,x,y)u(0,y)dy \\
&\qquad\quad + \int_0^t \frac{1}{2}\partial_y H_D(t-s,x,0)u(s,0)\, ds\ .
\end{aligned} \tag{3.12}$$

Initial data $u(0,x) = f(x)$, boundary data $u(t,0) = g(t)$, and inhomogeneous data $h(t,x)$ gives a full solution formula:

$$\begin{aligned}
u(t,x) = &\int_0^{+\infty} H_D(t,x,y)f(y)dy \\
&+ \int_0^t \frac{1}{2}\partial_y H_D(t-s,x,0)g(s)\, ds \\
&+ \int_0^t \int_0^{+\infty} H_D(t,x,y)h(s,y)\, dyds\ .
\end{aligned}$$

A similar line of reasoning gives a formula for solutions of the heat equation on $\mathbb{R}^1_+$ with prescribed inhomogeneous Neumann data on the boundary $x = 0$. Namely, given initial data $u(0,x) = f(x)$, boundary data $\partial_x u(t,0) = g(t)$, and inhomogeneous data $h(t,x)$, one finds the following:

$$\begin{aligned}
u(t,x) = &\int_0^{+\infty} H_N(t,x,y)f(y)dy \\
&- \int_0^t \frac{1}{2} H_N(t-s,x,0)g(s)\, ds \\
&+ \int_0^t \int_0^{+\infty} H_N(t,x,y)h(s,y)\, dyds\ .
\end{aligned} \tag{3.13}$$

3.5. Conservation laws and the evolution of moments

The heat equation is often called the *diffusion equation*, and indeed the physical interpretation of a solution is of a heat distribution or a particle density distribution that is evolving in time according to equation (3.1). That is, in probabilistic terms, if the distribution of outcomes of a random event at time t is described by the function $u(t,x)$, the quantity

$$P_t[a,b) = \int_a^b u(t,x)\, dx$$

represents the probability of the outcome of a random event taking its value in the interval $[a,b)$ at that time. For instance $P_t[a,b)$ might represent the probability that a random particle lies in the interval at that time.

For this interpretation we should ask that the initial data $f(x)$ satisfy $f(x) \geq 0$ and normalize it so that $\int_{\mathbb{R}^1} f(x)\, dx = 1$. In particular we require that $f \in L^1(\mathbb{R}^1)$. We can derive the heat equation itself from this interpretation given one more piece of information, namely Fourier's law of heat conservation and flux. Taking $[a, b)$ to be an arbitrary interval, consider the change in $u(t, x)$ in the interval over time. Fourier's law of heat conservation posits that this change is given by the flux of heat across the boundary. Namely, the change in heat over this interval in time is given by

$$\partial_t P_t[a, b) = \partial_t \int_a^b u(t, x)\, dx = F(a) - F(b) ,$$

where $F(a)$ is interpreted to be the flux of heat across the boundary point a into the interval, and $F(b)$ is interpreted to be the flux of heat out of the interval across the boundary point b. Fourier's law for the heat flux gives the form as

$$F(a) := -\frac{1}{2}\partial_x u(t, a) ,$$

and similarly for b so that

$$\begin{aligned} \partial_t P_t[a, b) &= \int_a^b \partial_t u(t, x)\, dx = F(a) - F(b) \\ &= \frac{1}{2}\partial_x u(t, b) - \frac{1}{2}\partial_x u(t, a) = \int_a^b \frac{1}{2}\partial_x^2 u(t, x)\, dx . \end{aligned}$$

Since the interval $[a, b)$ is arbitrary, the heat density $u(t, x)$ must satisfy the heat equation $\partial_t u = \frac{1}{2}\partial_x^2 u$.

This interpretation must satisfy two conditions in order to be consistent: (i) the total amount of heat must be conserved, and (ii) if the initial heat distribution $f(x) \geq 0$, then for all $t \geq 0$ we should have $u(t, x) \geq 0$. The second condition is already satisfied for us by the maximum principle. Regarding the first, there is the following result.

Theorem 3.8. *If $u(t, x) \in L^1(\mathbb{R}^1)$ is a solution of (3.1), then for all $t \geq 0$*

$$\int_{-\infty}^{+\infty} u(t, x)\, dx = \int_{-\infty}^{+\infty} f(x)\, dx .$$

Proof.

$$\begin{aligned} \partial_t \int_{-\infty}^{+\infty} u(t, x)\, dx &= \int_{-\infty}^{+\infty} \partial_t u(t, x)\, dx \\ &= \int_{-\infty}^{+\infty} \frac{1}{2}\partial_x^2 u(t, x)\, dx = \frac{1}{2} \lim_{R \to +\infty} \big(\partial_x u(t, R) - \partial_x u(t, -R)\big) = 0 . \end{aligned}$$

□

In fact the proof is incomplete, for we should also have shown that when $f(x) \in L^1(\mathbb{R}^1)$, then for $t > 0$ we have $\partial_x^2 u(t,x) \in L^1(\mathbb{R}^1)$, and then justified the exchanges of differentiation and integration and the integrations by parts. These facts will be addressed in the later chapter on the properties of the Fourier transform.

The interpretation of a solution $u(t,x)$ of (3.1) as the density of a probability measure motivates a discussion of the time evolution of its moments. For a given distribution $f(x)\,dx$, define the kth moment to be the quantity

$$m_k(f) = \int_{-\infty}^{+\infty} x^k f(x)\,dx \ . \tag{3.14}$$

For an arbitrary $f(x) \in L^1(\mathbb{R}^1)$ this may be infinite or not well defined, but for sufficiently well-localized functions it makes perfectly good sense. To give examples, take the case of the heat kernel itself, then

$$m_0(H(t,\cdot)) = \int_{-\infty}^{+\infty} \frac{1}{\sqrt{2\pi t}} e^{-\frac{x^2}{2t}}\,dx = 1 \ ,$$
$$m_1(H(t,\cdot)) = \int_{-\infty}^{+\infty} \frac{1}{\sqrt{2\pi t}} x e^{-\frac{x^2}{2t}}\,dx = 0 \ ,$$
$$m_2(H(t,\cdot)) = \int_{-\infty}^{+\infty} \frac{1}{\sqrt{2\pi t}} x^2 e^{-\frac{x^2}{2t}}\,dx = t \ .$$

Now consider the initial value problem for the heat equation; our solution procedure yields $u(t,x) = H(t,\cdot) * f(x)$, and the resulting heat distribution $u(t,x)\,dx$ has moments which are given by

$$m_k(u(t,\cdot)) = \int_{-\infty}^{+\infty} x^k u(t,x)\,dx = \iint x^k \frac{1}{\sqrt{2\pi t}} e^{-\frac{(x-y)^2}{2t}} f(y)\,dydx \ .$$

Proposition 3.9. *The first two moments of the solution $u(t,x)$ are conserved:*

$$m_0(u(t,\cdot)) = m_0(f) \ , \qquad m_1(u(t,\cdot)) = m_1(f) \ .$$

To deduce the evolution properties of the higher moments of $u(t,x)\,dx$, we will derive further elementary properties of the Fourier transform.

Proposition 3.10. *For sufficiently well-behaved functions $g(x)$, the Fourier transform has the property that*

$$\mathcal{F}(x^k g(x)) = (i\partial_\xi)^k \hat{g}(\xi) \ ,$$
$$\int_{-\infty}^{+\infty} g(x)\,dx = \sqrt{2\pi}\hat{g}(0) \ .$$

Proof. This first statement is essentially dual to that of Proposition 2.3. Observe by integrations by parts that

$$\int e^{-i\xi x}\big(x^k g(x)\big)\,dx = \int \big((i\partial_\xi)^k e^{-i\xi x}\big) g(x)\,dx = (i\partial_\xi)^k \int e^{-i\xi x} g(x)\,dx \ ,$$

which is the first statement of the proposition. To show the second statement,

$$\int g(x)\,dx = \int e^{-i\xi x} g(x)\,dx\big|_{\xi=0} = \sqrt{2\pi}\hat{g}(0) \ .$$

Hypotheses on the function $g(x)$ are necessary in order to justify the integrations by parts and the limits. □

We use this information in order to compute moments; since

$$\widehat{x^k u}(t,\xi) = (i\partial_\xi)^k \hat{u}(t,\xi) = \sqrt{2\pi}(i\partial_\xi)^k \big(\hat{H}(t,\xi)\hat{f}(\xi)\big) \ ,$$

therefore

$$\begin{aligned} m_k(u(t,\cdot)) &= \int x^k u(t,x)\,dx = \sqrt{2\pi}\big((i\partial_\xi)^k \hat{u}(t,\xi)\big)\big|_{\xi=0} \\ &= 2\pi\big((i\partial_\xi)^k \hat{H}(t,\xi)\hat{f}(\xi)\big)\big|_{\xi=0} \ . \end{aligned}$$

In particular for $k = 0$,

$$m_0(u(t,\cdot)) = 2\pi\hat{H}(t,0)\hat{f}(0) = \sqrt{2\pi}\hat{f}(0) = m_0(f) \ ,$$

which gives the first statement of Proposition 3.9. For $k = 1$,

$$\begin{aligned} m_1(u(t,\cdot)) &= 2\pi i\partial_\xi\big(\hat{H}(t,\xi)\hat{f}(\xi)\big)\big|_{\xi=0} \\ &= 2\pi i\big(\partial_\xi \hat{H}(t,\xi)\hat{f}(\xi) + \hat{H}(t,\xi)\partial_\xi \hat{f}(\xi)\big)\big|_{\xi=0} \\ &= 2\pi i\hat{H}(t,0)\partial_\xi \hat{f}(\xi)\big|_{\xi=0} = m_1(f) \ . \end{aligned}$$

This proves the second statement of Proposition 3.9, for initial data with finite zeroth and first moments, at least. The general case is now clear:

$$\begin{aligned} m_k(u(t,\cdot)) &= 2\pi(i\partial_\xi)^k\big(\hat{H}(t,\xi)\hat{f}(\xi)\big)|_{\xi=0} \\ &= 2\pi i^k \sum_{j=0}^{k} \binom{k}{j} \big(\partial_\xi^j \hat{H}(t,\xi)\partial_\xi^{k-j}\hat{f}(\xi)\big)|_{\xi=0} \\ &= \sum_{\substack{j=0 \\ j \text{ even}}}^{k} \binom{k}{j} m_j(H(t,\cdot))m_{k-j}(f) \ . \end{aligned}$$

We used the Leibnitz product rule for differentiation in the second line, and in the third line we used the fact that odd moments of the heat kernel are zero, as the kernel itself is an even function of x.

A property of solutions of the heat equation, related to that of conservation of heat as in Theorem 3.8, is that the evolution has a certain contraction property on a number of common spaces of functions.

Proposition 3.11. *Let $u(t,x)$ solve the heat equation (3.1) with initial data $f \in L^2(\mathbb{R}^1)$. Then the L^2-norm of $u(t,\cdot)$ is a decreasing function of t:*

$$\|u(t,\cdot)\|_{L^2} \le \|f\|_{L^2} \ .$$

Proof. Compute the time derivative of the norm:

$$\begin{aligned}\partial_t \|u(t,\cdot)\|_{L^2}^2 &= \partial_t \int u^2(t,x)\,dx \\ &= \int 2u(t,x)\partial_t u(t,x)\,dx = \int 2u(t,x)\frac{1}{2}\partial_x^2 u(t,x)\,dx \\ &= -\int |\nabla u(t,x)|^2\,dx \le 0 \ .\end{aligned}$$

Thus we find that

$$\|u(t,\cdot)\|_{L^2}^2 - \|f\|_{L^2}^2 = \int_0^t \partial_t \|u(s,\cdot)\|_{L^2}^2\,ds \le 0 \ .$$

One needs to revisit this calculation in order to justify the integration by parts. □

We have already seen that the setting of $L^1(\mathbb{R}^1)$ functions is sometimes more natural for the heat equation than that of $L^2(\mathbb{R}^1)$. It turns out that time evolution by heat flow is also a contraction in $L^1(\mathbb{R}^1)$.

Proposition 3.12. *Let $f \in L^1(\mathbb{R}^1)$, and let $u(t,x)$ solve the heat equation with initial data f. Then $\|u(t,\cdot)\|_{L^1}$ is a nonincreasing function of t:*

$$\|u(t,\cdot)\|_{L^1} \le \|f\|_{L^1} \ .$$

Proof. By nonincreasing we mean that the norm in question could stay the same or it could decrease in time; in any case it cannot increase. The easy case is for $f \in L^1$ to be of one sign; say for convenience that $f(x) \ge 0$. Then the maximum principle implies that $u(t,x) \ge 0$ as well, so that

$$\begin{aligned}\partial_t \int |u(t,x)|\,dx &= \partial_t \int u(t,x)\,dx \\ &= \int \partial_t\, u(t,x)\,dx = \int \frac{1}{2}\partial_x^2 u(t,x)\,dx = 0 \ ,\end{aligned}$$

and the L^1-norm is unchanged by the flow. However, when the initial data $f(x)$ changes sign, the L^1-norm will in fact be decreasing in time. To show this, decompose the initial data $f(x) = f_+(x) - f_-(x)$ into, respectively, its positive and negative parts. Both $f_\pm(x) \ge 0$, and furthermore they have disjoint support; $\{f_+ \neq 0\} \cap \{f_- \neq 0\} = \emptyset$. The L^1-norm of f is then the sum of the L^1-norms; indeed,

$$\int |f(x)|\,dx = \int (f_+(x) + f_-(x))\,dx = \int_{\mathrm{supp}(f_+)} f_+\,dx + \int_{\mathrm{supp}(f_-)} f_-\,dx\ .$$

The previous result can now be applied to the two initial data, f_+ and f_-, respectively:

$$\|u_+(t,\cdot)\|_{L^1} = \|f_+\|_{L^1}\ , \qquad \|u_-(t,\cdot)\|_{L^1} = \|f_-\|_{L^1}\ .$$

The solution is obtained from the sum, $u(t,x) = u_+(t,x) - u_-(t,x)$; however,

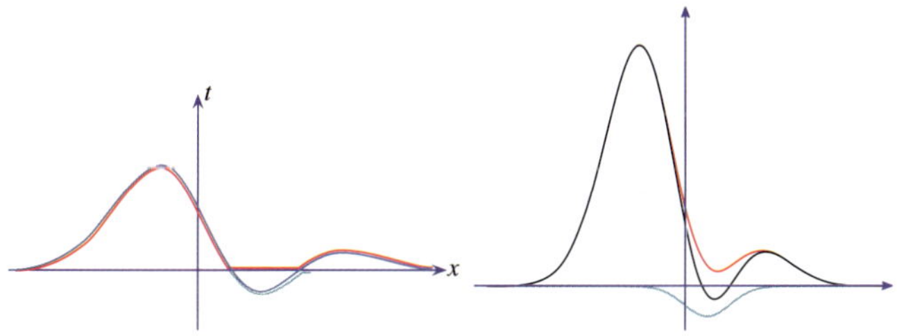

Figure 2. Decomposition of initial data into positive $(f_+(x))$ and negative $(-f_-(x))$ components, and their subsequent evolution under heat flow, respectively $u_+(t,x)$ and $u_-(t,x)$.

the maximum principle applies so that as long as neither of $f_\pm$ is zero almost everywhere, $\mathrm{supp}(u_+) = \mathrm{supp}(u_-) = \mathbb{R}^1$. This decomposition of $f(x)$ and its time evolution is illustrated in Figure 2. We therefore have the estimate

$$\begin{aligned}\|u(t,\cdot)\|_{L^1} &= \|u_+ - u_-\|_{L^1} \\ &\leq \|u_+\|_{L^1} + \|u_-\|_{L^1} = \|f_+\|_{L^1} + \|f_-\|_{L^1} = \|f\|_{L^1}\ .\end{aligned}$$

Since we are assuming that both u_+ and u_- are not zero, there must be some cancellation in the inequality of the second line, and hence the inequality is in fact strict. □

3.6. The heat equation in $\mathbb{R}^n$

Most of the results of this chapter generally hold for the heat equation posed in higher-dimensional space $\mathbb{R}^n$ for any dimension n. The heat equation is

$$\begin{aligned}\partial_t u &= \frac{1}{2}\Delta u\ , \qquad t \in \mathbb{R}^1_+\ , \quad x \in \mathbb{R}^n\ , \\ u(0,x) &= f(x)\ .\end{aligned} \tag{3.15}$$

For initial data $f(x)$, we follow the procedure of section 3.1 and use the Fourier transform, this time in $\mathbb{R}^n$:

$$\hat{f}(\xi) = \frac{1}{\sqrt{2\pi}^n} \int_{\mathbb{R}^n} e^{-i\xi\cdot y} f(y)\,dy := \mathcal{F}f(\xi),$$

where both ξ and y are in $\mathbb{R}^n$ and where their Euclidean inner product is given by $\xi \cdot y = \sum_{j=1}^{n} \xi_j y_j$. The inverse transform is similarly given by

$$f(x) = \frac{1}{\sqrt{2\pi}^n} \int_{\mathbb{R}^n} e^{i\xi\cdot x} \hat{f}(\xi)\, d\xi := \mathcal{F}^{-1}\hat{f}(x) ,$$

and the Plancherel identity holds for $f(x) \in L^2(\mathbb{R}^n)$:

$$\|f\|^2_{L^2(\mathbb{R}^n_x)} = \|\hat{f}\|^2_{L^2(\mathbb{R}^n_\xi)} . \tag{3.16}$$

A solution $u(t,x)$ of (3.15) has Fourier transform that satisfies the parametrized family of ODEs

$$\frac{d}{dt}\hat{u} + \frac{1}{2}|\xi|^2\hat{u} = 0$$

which has solution $\hat{u}(t,\xi) = e^{-\frac{1}{2}|\xi|^2 t}\hat{f}(\xi)$, analogous to our previous expression. As we did in (3.2), the heat kernel is given by a Gaussian integral

$$H(t, x-y) = \frac{1}{(2\pi t)^n} \int_{\mathbb{R}^n} e^{i\xi\cdot(x-y)} e^{-\frac{1}{2}|\xi|^2 t}\, d\xi = \frac{1}{\sqrt{2\pi t}^n} e^{-\frac{|x-y|^2}{2t}} . \tag{3.17}$$

Thus a solution to (3.15) is given by convolution with the heat kernel

$$u(t,x) = \int_{\mathbb{R}^n} H(t, x-y) f(y)\, dy . \tag{3.18}$$

The heat kernel gives rise to a probability measure that is important for Brownian motion in $\mathbb{R}^n$,

$$dP_t(x) = \frac{1}{\sqrt{2\pi t}^n} e^{-\frac{|x|^2}{2t}}\, dx ,$$

which is invariant under Brownian scaling and has total measure $\int dP_t(x) = 1$, as one can check.

3.7. Entropy

The interpretation of nonnegative and normalized solutions of the heat equation as probability distributions has been discussed in section 3.5. Associated with such probability measures are a variety of concepts that concern the *entropy* of a measure. Given a density $f \in L^1(\mathbb{R}^n)$ with $f(x) \geq 0$ and $\|f\|_{L^1} = 1$, the entropy is defined as

$$S(f) := \int_{\mathbb{R}^n} f(x) \log(f(x))\, dx \tag{3.19}$$

with the entropy density given by

$$s(x) = f(x)\log(f(x)) . \tag{3.20}$$

The concept of entropy dates to the work of Boltzmann, who recognized the quantity as a measure of the disorder of a distribution of particles in the context of the dynamics of gasses. The definition above is made as an analogy to this idea when applied to heat flow. There are two fundamental physical measurements when considering heat;[1] one is the quantity of heat, given by

$$\int_{\mathbb{R}^n} u(t,x)\,dx = m_0(u(t,\cdot))\ ,$$

which is conserved as described in Proposition 3.9. The second is a measure of the amount of disorder in the heat distribution, given by the entropy $S(u(t,\cdot))$. Work can be performed by ordered distributions of heat, as in, for example, a steam engine. Entropy decreases the amount of work that can be performed for the same quantity of heat. If $f(x)$ is taken as initial data for the heat equation, the entropy of the solution $u(t,x)$ evolves monotonically in time; this is the conclusion of the following proposition.

Proposition 3.13. *Suppose that $f(x) \geq 0$ is taken as initial data for the heat equation. Then the solution $u(t,x)$ given by* (3.18) *satisfies*

$$\frac{d}{dt} S(u(t,\cdot)) \leq 0\ .$$

Proof. Unless $f(x) = 0$, we know by the maximum principle that $u(t,x) > 0$, and hence the entropy density $s(t,x)$ is at least defined. The time evolution of the total entropy density in a domain $D \subseteq \mathbb{R}^n$ can be expressed by

$$\begin{aligned}\frac{d}{dt}\int_D s(t,x)\,dx &= \frac{d}{dt}\int_D u(t,x)\log(u(t,x))\,dx \\ &= \int_D \partial_t u\,\log(u)\,dx + \int_D u\,\frac{\partial_t u}{u}\,dx \\ &= \int_D \frac{1}{2}\Delta u\,\log(u)\,dx + \int_D \frac{1}{2}\Delta u\,dx\ .\end{aligned}$$

From integrations by parts, this equals

$$\begin{aligned}&= \int_{\partial D} \frac{1}{2}(\log(u)+1)\nabla u\cdot N\,dS - \frac{1}{2}\int_D \nabla u\cdot\nabla\log(u)\,dx \\ &= \int_{\partial D} \frac{1}{2}(\log(u)+1)\nabla u\cdot N\,dS - \frac{1}{2}\int_D \frac{|\nabla u|^2}{u}\,dx\ .\end{aligned}$$

Taking the limit of large domains D which exhaust all of $\mathbb{R}^n$, for example the limit of large balls $D = B_R(0)$ as $R \to +\infty$, the conclusion is that

$$\frac{d}{dt}\int_{\mathbb{R}^n} u(t,x)\log(u(t,x))\,dx = -\frac{1}{2}\int_{\mathbb{R}^n}\frac{|\nabla u|^2}{u}\,dx \leq 0\ .$$

Therefore $S(u(t,\cdot))$ is a nonincreasing quantity associated with heat flow. □

[1] Freeman Dyson, *Scientific American*, September 1954.

In the proof one identifies the rate of change of entropy as being

$$-\frac{1}{2}\int_{\mathbb{R}^n} \frac{|\nabla u|^2}{u}\, dx \ , \tag{3.21}$$

and the integration by parts also identifies the *entropy flux* to be

$$F(u) := \frac{1}{2}\big(1+\log(u)\big)\nabla u \ .$$

This shows that the quantity $S(u(t,\cdot))$ is decreasing; by mathematics tradition we have evidently chosen the opposite sign in definition (3.19) from that in practice in physics, where *entropy* is defined to be $-S(u)$ and is increasing. The entropy $S(f)$ is not guaranteed to be finite for any choice of $f \in L^1(\mathbb{R}^n)$, but if it is initially so, then it will remain finite for $u(t,\cdot)$ for all future $t > 0$.

There are other quantities that are studied in relation to the concept of entropy. One is the *relative entropy* of one probability distribution with respect to another. Suppose that $u(x)$ and $e(x)$ are densities of two probability measures; the relative entropy of $dP_1(x) = u(x)dx$ with respect to $dP_2(x) = e(x)dx$ is defined to be

$$S(u|e) := \int_{\mathbb{R}^n} u(x)\log\Big(\frac{u(x)}{e(x)}\Big)\, dx \ .$$

Note that there is an asymmetry between the roles of u and e. Suppose that $u(t,x)$ is a solution of the heat equation and that $e(t,x) = H(t,x)$ is given by the heat kernel (3.17). The change in time of the relative entropy is then expressed as

$$\frac{d}{dt}\int_{\mathbb{R}^n} u\log\frac{u}{e}\, dx = \int_{\mathbb{R}^n} \partial_t\big(u\log(u) - u\log(e)\big)\, dx \ .$$

The first term of the RHS we know from (3.21). The second term evolves as follows:

$$\begin{aligned}\frac{d}{dt}\int_{\mathbb{R}^n} u\log(e)\, dx &= \int_{\mathbb{R}^n} \partial_t u\log(e) + u\frac{\partial_t e}{e}\, dx = \int_{\mathbb{R}^n} \frac{1}{2}\Delta u\log(e) + u\frac{\frac{1}{2}\Delta e}{e}\, dx \\ &= -\int_{\mathbb{R}^n} \frac{1}{2}\nabla u\cdot\frac{\nabla e}{e} + \frac{1}{2}\nabla\Big(\frac{u}{e}\Big)\cdot\nabla e\, dx = -\int_{\mathbb{R}^n} \nabla u\cdot\frac{\nabla e}{e} - \frac{1}{2}u\frac{|\nabla e|^2}{e^2}\, dx \ .\end{aligned}$$

Since $e(t,x) = H(t,x)$, the heat kernel, the latter quantity is explicit:

$$\frac{\nabla e}{e} = -\frac{x}{t} \ .$$

We can therefore compute that

$$\int_{\mathbb{R}^n} \nabla u\cdot\frac{\nabla e}{e}\, dx = -\int_{\mathbb{R}^n} \nabla u\cdot\frac{x}{t}\, dx = \frac{n}{t}\int_{\mathbb{R}^n} u\, dx \ ,$$

and furthermore

$$\int_{\mathbb{R}^n} -\frac{1}{2}u\frac{|\nabla e|^2}{e^2}\,dx = -\int_{\mathbb{R}^n} \frac{1}{2}u\frac{|x|^2}{t^2}\,dx = -\frac{1}{2t^2}\int_{\mathbb{R}^n} |x|^2 u(t,x)\,dx \ .$$

The latter is the variance of the probability distribution $dP_1 = u(t,x)dx$, which from section 3.5 we know satisfies

$$\partial_t \int_{\mathbb{R}^n} |x|^2 u\,dx = \frac{1}{2}\int_{\mathbb{R}^n} |x|^2\Delta u\,dx = \frac{1}{2}\int_{\mathbb{R}^n} \Delta(|x|^2)\,u\,dx = n\int_{\mathbb{R}^n} u(t,x)\,dx \ .$$

Since by Proposition 3.9 the moment $m_0(u(t,\cdot)) = m_0(f)$ is a constant, a conserved quantity, we then have that

$$\begin{aligned} -\int_{\mathbb{R}^n} \nabla u\cdot\frac{\nabla e}{e} - \frac{1}{2}u\frac{|\nabla e|^2}{e^2}\,dx \\ = -\frac{n}{t}\int_{\mathbb{R}^n} u\,dx + \Big(\frac{n}{2t}\int_{\mathbb{R}^n} u\,dx + \frac{1}{2t^2}\int_{\mathbb{R}^n} |x|^2 f(x)\,dx\Big) \ . \end{aligned}$$

Putting these together, we have a quite explicit form for the evolution of the relative entropy $S(u|e)$, namely

$$\frac{d}{dt}S(u|e) = -\frac{1}{2}\int_{\mathbb{R}^n} \frac{|\nabla u|^2}{u}\,dx - \frac{n}{2t}m_0(f) - \frac{1}{t^2}m_2(f) \ . \tag{3.22}$$

3.8. Gradient flow

This section of the chapter has to do with an interpretation of the heat equations as a gradient flow for the *energy functional*

$$E(u) = \frac{1}{4}\int |\nabla u|^2\,dx \ . \tag{3.23}$$

To illustrate gradient flow with a finite-dimensional example, let

$$e(v)\ :\ \mathbb{R}^d \to \mathbb{R}$$

be a C^1 function of $v \in \mathbb{R}^d$. Its gradient $\nabla e(v)$ (defined in terms of the Euclidean inner product on $\mathbb{R}^d$) is given by the formula

$$\frac{d}{d\sigma}\Big|_{\sigma=0} e(v+\sigma V) = \langle \nabla e(v), V\rangle \ .$$

The (negative) gradient flow of $e(v)$ is the solution map of the ODE

$$\dot v = -\nabla e(v) \ .$$

A basic property of gradient flow is that the function $e(v)$ is decreasing along trajectories; indeed,

$$\frac{d}{dt}e(v(t)) = \langle \nabla e(v), \dot v(t)\rangle = -\langle \nabla e(v), \nabla e(v)\rangle = -\|\nabla e(v)\|^2 \le 0 \ .$$

Such flows play a role in Morse theory, among many other things.

The heat equation defines a flow in space of functions. Using a similar calculation, we can identify this as a gradient flow for the energy functional $E(u)$:

$$\begin{aligned}\frac{d}{d\sigma}\Big|_{\sigma=0} E(u+\sigma w) &= \frac{d}{d\sigma}\Big|_{\sigma=0} \frac{1}{4}\int |\nabla u + \sigma w|^2 \, dx \\ &= \frac{d}{d\sigma}\Big|_{\sigma=0} \Big(\frac{1}{4}\int |\nabla u|^2 + 2\sigma \nabla u \cdot \nabla w + \sigma^2 |\nabla w|^2 \, dx\Big) \\ &= \frac{1}{2}\int \nabla u \cdot \nabla w \, dx \ .\end{aligned}$$

In order to describe this quantity in terms of the L^2 inner product, one integrates by parts:

$$\frac{d}{d\sigma}\Big|_{\sigma=0} E(u+\sigma w) = -\int \frac{1}{2}\Delta u \, w \, dx \ . \tag{3.24}$$

This identifies $\text{grad} E(u) = -\frac{1}{2}\Delta u$. The (negative) gradient flow is therefore

$$\partial_t u = -\text{grad} E(u) = \frac{1}{2}\Delta u \ ,$$

which is precisely the heat equation (3.15). A classical notation is to write $\delta E(u)$ for the *Gâteau derivative* of $E(u)$; that is, for the directional derivative of $E(u)$ in the direction w, one writes that

$$\frac{d}{d\sigma}\Big|_{\sigma=0} E(u+\sigma w) = \langle \delta E(u), w\rangle \ .$$

Exercises: Chapter 3

Exercise 3.1. This problem concerns the semigroup property of the solution process for the heat equation. Show that the heat kernel satisfies the identity

$$H(t,x) = \int_{-\infty}^{+\infty} H(t-s, x-y) H(y,s) \, dy \quad \forall 0 < s < t \ .$$

Now show that the heat operator satisfies the property that for all $0 < s < t$,

$$\mathbf{H}(t) = \mathbf{H}(t-s)\mathbf{H}(s) \ .$$

Conclude that formula (3.5) holds for the one-parameter family of solution operators $\{\mathbf{H}(t)\}_{t\in\mathbb{R}^1_+}$ of the heat equation.

Exercise 3.2. Justify on a rigorous level of analysis the exchange of integrations in the proof of Proposition 3.4(iii), therefore completing the rigorous proof of the proposition's three parts.

Exercise 3.3. Solve the following initial value problems for the heat equation in explicit terms.

(1) $f(x) = x$

(2) $f(x) = x^2$

(3) $f(x) = 0$ for $x < 0$ and $f(x) = 1$ for $x \geq 0$

(4) $f(x) = e^{\alpha x}$

(5) $f(x) = \sin(kx)$

What is the asymptotic behavior of $u(t,x)$ as $t \to +\infty$? What is their asymptotic behavior as $t \to +\infty$? On what properties of the initial data does this behavior depend? Does it affect the asymptotic behavior if $f(x) \notin L^1(\mathbb{R}^1)$?

Exercise 3.4 (Moments of the heat kernel)**.** (i) Show that the second moment of the heat kernel

$$H(t,x) = \frac{1}{\sqrt{2\pi t}} e^{-\frac{x^2}{2t}} = P_t(x)$$

is its variance and that

$$m_2(P_t) = t \ .$$

Hint: Calculate the following integral:

$$G(\alpha) = \frac{1}{\sqrt{2\pi}} \int_{-\infty}^{+\infty} e^{-\alpha \frac{x^2}{2}} \, dx \ ,$$

and then use the fact that

$$\frac{1}{\sqrt{2\pi}} \int_{-\infty}^{+\infty} e^{-\frac{x^2}{2}} x^2 \, dx = -2\partial_\alpha G(1) \ .$$

(ii) Also calculate all the moments $m_j(P_t)$ for j any odd integer.

(iii) Using the same technique as in (i), give an expression for all of the moments $m_j(P_t)$ for j even.

Exercise 3.5. Suppose that the initial data $f(x) \in L^1(\mathbb{R}^1)$ for the heat equation is piecewise continuous, meaning that it contains a possibly finite number of jump discontinuities

$$\lim_{x \to x_j^-} f(x) = a_j \ , \qquad \lim_{x \to x_j^+} f(x) = b_j$$

at points x_j, $j = 1, N$. Show that the limit of the solution (3.4) of the heat equation satisfies

$$\lim_{t \to 0^+} u(t, x_j) = \frac{1}{2}(a_j + b_j) \ .$$

Exercise 3.6 (Method of images)**.** Give an expression for the solution of the heat equation over the bounded interval $0 \leq x < 2\pi$ for the following cases of boundary conditions.

(1) Dirichlet: $u(t,0) = 0 = u(t,2\pi)$. Denote this solution by $u_D(t,x)$.

(2) Neumann: $-\partial_x u(t,0) = 0 = \partial_x u(t,2\pi)$. Denote this solution by $u_N(t,x)$.

(3) Periodic: $u(t,x) = u(t,x+2\pi)$. Denote this solution by $u_p(t,x)$.

Suppose that the initial data satisfies $0 \le f(x) \le 1$ for the heat equation on $[0,2\pi]$ (and assume that f is not identically zero). Show that in each case $u = u_D$, u_N, and u_p, and for all $t > 0$ the solution satisfies the inequality

$$0 < u(t,x) < 1$$

(unless $f(x) \equiv 0$ or $f(x) \equiv 1$, in which case $u(t,x) \equiv 0$ or $u(t,x) \equiv 1$, respectively).

Exercise 3.7. (i) Using the method of separation of variables, give an expression for the solution $u(t,x)$ for $t > 0$ to the heat equation with Dirichlet boundary conditions

$$u(t,0) = 0 = u(t,2\pi) \ ,$$

given the initial conditions

$$u(0,x) = f(x) = x \ .$$

(ii) Clearly $u(0,2\pi) \ne 0$. Show that for all $t > 0$ the solution nevertheless satisfies $u(t,2\pi) = 0$.

Exercise 3.8. Consider the three solutions with initial data $0 \le f(x)$ for the three different boundary conditions in Exercise 3.6. Show that for all $0 \le x \le 2\pi$ and all $t > 0$, the three solutions satisfy

$$u_D(t,x) < u_p(t,x) \quad \text{and} \quad u_D(t,x) < u_N(t,x) \ .$$

Is it also true that

$$u_p(t,x) < u_N(t,x) \ ?$$

Exercise 3.9. Suppose that the initial data $f(x) \in L^1(\mathbb{R}^n)$, and consider the solution $u(t,x)$ given by convolution with the heat kernel. Show that

$$\lim_{t\to+\infty} u(t,x) = 0$$

uniformly in $x \in \mathbb{R}^n$. Furthermore, give a decay rate for the solution by showing that it is bounded by

$$|u(t,x)| \le \frac{1}{\sqrt{2\pi t}^{\,n}} \|f\|_{L^1(\mathbb{R}^n)} \ .$$

Exercise 3.10 (Entropy). (i) Define two distributions by their densities:

$$f_1(x) = \chi_{-1/2,+1/2}(x) \ , \qquad f_2(x) = 2\chi_{-1/4,+1/4} \ .$$

They both have total mass $m_0(f_j) = 1$ for $j = 1,2$. Compare their entropies. Which is larger (less disordered)?

(ii) What is the entropy of the heat kernel (3.3)

$$H(t,x) = \frac{1}{\sqrt{2\pi t}} e^{-\frac{x^2}{2t}}\, dx \ ?$$

How does it behave as $t \to 0+$ and as $t \to +\infty$?

Exercise 3.11. Suppose the functions $f_1(x) \geq 0$ and $f_2(x) \geq 0$ have disjoint support, meaning that $\{x : f_1(x) > 0\} \cap \{x : f_2(x) > 0\} = \emptyset$.

(i) Show that the entropy of the function $f(x) = f_1(x) + f_2(x)$,

$$S(f) = \int_{-\infty}^{+\infty} f(x) \log(f(x))\, dx \ ,$$

satisfies

$$S(f) = S(f_1) + S(f_2) \ .$$

(ii) Now suppose that $u_1(t,x)$ and $u_2(t,x)$ are the solutions of the heat equation with initial data f_1 and f_2, respectively. Let $u(t,x) = u_1(t,x) + u_2(t,x)$. Show that

$$S(u(t,\cdot)) \geq S(u_1(t,\cdot)) + S(u_2(t,\cdot)) \ .$$

Under what condition is this inequality an equality?

Projects: Chapter 3

Project 3.12 (Tikhonov-nonuniqueness of solutions of the heat equation). It turns out that solutions of the heat equations are not unique, at least if one admits solutions that grow very rapidly and oscillate as $|x| \to +\infty$. This is a surprising result that appeared in Tikhonov [**20**]. To show this we will solve the heat equation with data on the positive t-axis in the form

$$\begin{aligned} &\partial_t u = \frac{1}{2}\partial_x^2 u \ , \quad x \in \mathbb{R}^1 \ , \\ &u(t,0) = g(t) \ , \ \partial_x u(t,0) = 0 \ , \quad t \in \mathbb{R}_+ \ . \end{aligned}$$

(i) Derive a formal power series expansion in x for a solution $u(t,x)$ in the form

$$u(t,x) = \sum_{j \geq 0} c_j x^j \ ,$$

and use the heat equation to show that for all $j \in \mathbb{N}$,

$$c_{2j} = \frac{2^j}{(2j)!}\partial_t^j g(t) \ , \quad c_{2j+1} = 0 \ .$$

(ii) Give functions $g(t)$ such that this formal series for $u(t,x)$ converges for all $x \in \mathbb{R}^1$. For example, functions in the *Gevrey class* will do, such as

$$g(t) = e^{-\frac{1}{t^2}} \ , \quad t > 0 \ .$$

Show that for this $g(t)$,

$$g(0) = 0 \ , \quad \partial_t^j g(0) = 0$$

and that for all $j \geq 0$,

$$|\partial_t^j g(t)| \leq \frac{(j!)^{3/2}}{(Ct)^j} e^{-\frac{1}{t^2}} \ .$$

Show that under these conditions the power series for $u(t, x)$ converges for $t > 0$. Furthermore show that

$$u(0, x) = 0 \ .$$

Such solutions are oscillatory and badly unbounded for large $x \in \mathbb{R}^1$. It turns out that if we impose growth bounds on $u(t, x)$, such as

$$|u(t, x)| \leq Ce^{x^2} \ ,$$

then solutions are unique in this class, meaning that $u(0, x) = 0$ implies $u(t, x) = 0$ for $t > 0$. Alternatively, if solutions are required to be bounded from below, then they are also unique.

Chapter 4

Laplace's equation

Soap bubbles: surfaces of constant mean curvature supported by a slight pressure difference across the interface. Image credit: Godunova Tatiana / iStock / Getty Images Plus.

4.1. Dirichlet, Poisson, and Neumann boundary value problems

The most commonly occurring form of problem that is associated with Laplace's equation is a boundary value problem, normally posed on a domain $\Omega \subseteq \mathbb{R}^n$. That is, Ω is an open set of $\mathbb{R}^n$ whose boundary is smooth enough so that integrations by parts may be performed; thus at the very least the boundary is rectifiable. The most common boundary value problem for Laplace's equation (1.6) is the *Dirichlet problem*

$$\begin{aligned} \Delta u(x) &= 0 , \qquad x \in \Omega , \\ u(x) &= f(x) , \qquad x \in \partial\Omega , \end{aligned} \tag{4.1}$$

where one specifies the values of the solution $u(x)$ on the boundary $\partial\Omega$, analogous to specifying the value of a solution of the wave or heat equation on the boundary of its domain. The function $f(x)$ is known as the Dirichlet data. In physics applications, $f(x)$ corresponds to a density distribution of dipoles of charged particles fixed on the boundary $\partial\Omega$, whereupon the solution $u(x)$ corresponds to the resulting electrostatic potential. A function satisfying $\Delta u = 0$ is called *harmonic*, as we have stated in Chapter 1.

Perhaps the second most common problem is called the *Poisson problem*,

$$\begin{aligned} \Delta u(x) &= h(x) , \qquad x \in \Omega , \\ u(x) &= 0 , \qquad x \in \partial\Omega , \end{aligned} \tag{4.2}$$

for which the function $h(x)$ represents a distribution of fixed charges in the domain Ω, while the boundary $\partial\Omega$ is a perfect conductor. Again the solution $u(x)$ represents the resulting electrostatic potential. There are several other quite common boundary value problems that are similar in character to (4.1), for example the *Neumann problem*

$$\begin{aligned} \Delta u(x) &= 0 , \qquad x \in \Omega , \\ \partial_N u(x) &= g(x) , \qquad x \in \partial\Omega , \end{aligned} \tag{4.3}$$

where $N(x)$ is the outward unit normal vector to Ω and $\partial_N u(x) = \nabla u(x) \cdot N$. The solution corresponds to the electrostatic potential in Ω due to a charge density distribution on $\partial\Omega$. The *Robin problem*, or boundary value problem of the third kind, asks to find $u(x)$ such that

$$\begin{aligned} \Delta u(x) &= 0 , \qquad x \in \Omega , \\ \partial_N u(x) - \beta u(x) &= g(x) , \qquad x \in \partial\Omega , \end{aligned} \tag{4.4}$$

where β is a real constant or possibly a real function of $x \in \Omega$. This problem is often associated with an imposed impedence on the boundary.

4.2. Green's identities

Consider two functions $u(x)$ and $v(x)$ defined on a domain $\Omega \subseteq \mathbb{R}^n$. Integration by parts gives the following formula, which is known as *Green's first identity*:

$$\int_\Omega v \Delta u \, dx = -\int_\Omega \nabla v \cdot \nabla u \, dx + \int_{\partial\Omega} v \partial_N u \, dS_x . \tag{4.5}$$

The notation is that dS_x is the differential of surface area in the integral over the boundary. To ensure that the manipulations in this formula are valid, we take $\Omega \subseteq \mathbb{R}^n$ to be bounded and ask that $u, v \in C^2(\Omega) \cap C^1(\overline{\Omega})$; that is, all derivatives of u and v up to second order are continuous in the interior of Ω and at least their first derivatives have continuous limits on the boundary $\partial\Omega$.

Integrating again by parts (or alternatively using (4.5) in a symmetric way with the roles of u and v reversed), we obtain *Green's second identity*

$$\int_\Omega v\Delta u\, dx - \int_\Omega \Delta v\, u\, dx = \int_{\partial\Omega} (v\partial_N u - \partial_N v u)\, dS_x \tag{4.6}$$

for $u, v \in C^2(\Omega) \cap C^1(\overline{\Omega})$. The integrand of the integral over the boundary $\partial\Omega$ is the analog of the Wronskian in ODEs.

Considering the function v as a test function and substituting several astute choices for it into Green's identities, we obtain information about solutions u of Laplace's equation. First of all, let $v(x) = 1$; then (4.5) gives

$$\int_\Omega \Delta u\, dx = \int_{\partial\Omega} \partial_N u\, dS_x \ .$$

In case u is harmonic, $\Delta u = 0$ and the LHS vanishes. This is a compatibility condition for boundary data $g(x) = \partial_N u$ for the Neumann problem.

Proposition 4.1. *In order for the Neumann problem* (4.3) *to have a solution, the Neumann data $g(x)$ must satisfy*

$$\int_{\partial\Omega} g(x)\, dS_x = \int_{\partial\Omega} \partial_N u(x)\, dS_x = 0 \ .$$

For a second choice, let $v(x) = u(x)$ itself. Then Green's first identity (4.5) is an identity involving the L^2-norm of ∇u, which is a form of an "energy" identity

$$\int_\Omega |\nabla u(x)|^2\, dx + \int_\Omega u\Delta u\, dx = \int_{\partial\Omega} u\partial_N u\, dS_x \ . \tag{4.7}$$

One consequence of this is a uniqueness theorem.

Theorem 4.2. *Suppose that $u \in C^2(\Omega) \cap C^1(\overline{\Omega})$ satisfies the Dirichlet problem* (4.1) *with $f(x) = 0$, the Poisson problem* (4.2) *with $h(x) = 0$, or the Neumann problem* (4.3) *with $g(x) = 0$. Then*

$$\int_\Omega |\nabla u(x)|^2\, dx = \int_{\partial\Omega} u\partial_N u\, dS_x = 0 \ .$$

Therefore $u(x) = 0$ in the cases of the Dirichlet problem (4.3) *and the Poisson problem* (4.2)*. In the case of the Neumann problem, the conclusion is that $u(x)$ is constant.*

Proof. Identity (4.7) implies that

$$\int_\Omega |\nabla u(x)|^2 \, dx = 0$$

in each of the three cases, which in turn implies that $\nabla u = 0$ almost everywhere in Ω. Therefore $u(x)$ must be constant as we have assumed that $u \in C^2(\Omega)$. In cases (4.1) and (4.2), this constant must vanish in order to satisfy the boundary conditions. In the case of the Neumann problem (4.3), we only conclude that $u(x)$ is constant. In this situation we conclude that the constant functions $u(x) \equiv C$ span the null space of the Laplace operator with Neumann boundary conditions. □

The case of unbounded domains Ω is not too different, and a uniqueness result is the goal. One asks that solutions $u(x)$ are bounded and have the limit $u(x) = 0$ as x becomes unbounded.

4.3. The fundamental solution

The Laplace operator is invariant under rotations, and one can imagine that solutions which are also rotationally invariant are of special interest. In polar coordinates (r, φ) in $\mathbb{R}^n$, where $r \in [0, +\infty)$ and $\varphi \in \mathbb{S}^{n-1}$, the Laplacian is expressed

$$\Delta u = \partial_r^2 + \frac{n-1}{r}\partial_r u + \frac{1}{r^2}\Delta_\varphi u \ , \tag{4.8}$$

where Δ_φ is the Laplace operator on the unit sphere $\mathbb{S}^{n-1} \subseteq \mathbb{R}^n$. A rotationally invariant solution $\Gamma(r)$ of Laplace's equation must satisfy

$$\partial_r^2 \Gamma + \frac{n-1}{r}\partial_r \Gamma = 0 \ ,$$

that is, $\partial_r \Gamma = \frac{C}{r^{n-1}}$, which in turn implies that

$$\Gamma(r) = \frac{C}{2-n}\frac{1}{r^{n-2}} \ , \qquad n \geq 3 \ ,$$
$$\Gamma(r) = C\log(r) \ , \qquad n = 2 \ .$$

The function $\Gamma(r)$ is harmonic for $0 < r < +\infty$ but singular for $r = 0$. It is called a *fundamental solution*, which is part of the content of Theorem 4.3. Its proof is the outcome of the following computation that uses Green's identities. Suppose that $u \in C^2(\Omega) \cap C^1(\partial\Omega)$ and consider $y \in \Omega$ a point inside the domain under consideration. Take $\rho > 0$ sufficiently small so that the ball $B_\rho(y) \subseteq \Omega$, as in Figure 1. Using Green's second identity (4.6) over

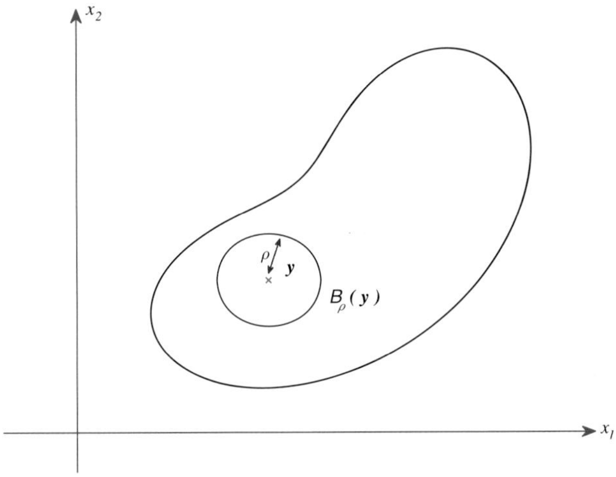

Figure 1. Domain Ω with a ball $B_\rho(y)$ excised.

the region $\Omega \backslash B_\rho(y)$, we have

$$\begin{aligned}
(4.9) \qquad & \int_{\Omega \backslash B_\rho(y)} \Gamma(|x-y|)\Delta u \, dx \\
&= \int_{\partial\Omega} (\Gamma \partial_N u - \partial_N \Gamma u) \, dS_x + \int_{S_\rho(y)} (\Gamma \partial_N u - \partial_N \Gamma u) \, dS_x \\
&+ \int_{\Omega \backslash B_\rho(y)} \Delta \Gamma \, u \, dx \ .
\end{aligned}$$

The fundamental solution Γ is harmonic in $\Omega \backslash B_\rho(y)$, the singularity at $x = y$ being inside $B_\rho(y)$; therefore that last term of the RHS is zero. The first term of the RHS is our usual boundary integral from (4.6). We are to calculate the second integral and its limit as $\rho \to 0$. On the sphere $S_\rho(y)$ we have $\Gamma(|x-y|) = \Gamma(\rho)$ while $\partial_N \Gamma(|x-y|) = -\partial_r \Gamma(\rho)$, the latter minus sign coming from the fact that the outward unit normal to $\Omega \backslash B_\rho(y)$ on $S_\rho(y)$ is pointing inwards towards y. Therefore, using the divergence theorem,

$$\int_{S_\rho(y)} \Gamma \partial_N u \, dS_x = \Gamma(\rho) \int_{S_\rho(y)} N \cdot \nabla u \, dS_x = -\Gamma(\rho) \int_{B_\rho(y)} \Delta u \, dx \ ,$$

and the limit of this quantity vanishes as $\rho \to 0$. Indeed, in case $n \geq 3$,

$$(4.10) \qquad \lim_{\rho \to 0} \left| \Gamma(\rho) \int_{B_\rho(y)} \Delta u \, dx \right| \leq \lim_{\rho \to 0} \frac{C}{n-2} \frac{1}{\rho^{n-2}} \rho^n \|u\|_{C^2}$$

which vanishes like ρ^2 as $\rho \to 0$. The case $n = 2$ also vanishes in the limit, which involves $\log(\rho)$ instead. On the other hand, the second term of (4.9) is

$$-\int_{S_\rho(y)} \partial_N \Gamma \, u \, dS_x = \frac{C(2-n)}{\rho^{n-1}} \int_{S_\rho(y)} u \, dS_x \ ,$$

and as ρ tends to zero, by the continuity of $u(x)$ at $x = y$, this has the limit

(4.11)
$$\lim_{\rho\to 0} \frac{C(2-n)}{\rho^{n-1}} \int_{S_\rho(y)} u\, dS_x = \lim_{\rho\to 0} \frac{C(2-n)}{\rho^{n-1}} \big(\omega_n \rho^{n-1} u(y)\big) = C(2-n)\omega_n u(y) ,$$

where the quantity ω_n is the surface area of the unit sphere $\mathbb{S}^{n-1} \subseteq \mathbb{R}^n$.

Theorem 4.3. *For $u \in C^2(\Omega) \cap C^1(\partial\Omega)$ and for $y \in \Omega$ we have the identity*

(4.12)
$$\int_\Omega \Gamma(|x-y|)\Delta u(x)\, dx = C(2-n)\omega_n u(y) + \int_{\partial\Omega} \big(\Gamma \partial_N u - \partial_N \Gamma\, u\big)\, dS_x .$$

Setting $C = ((2-n)\omega_n)^{-1}$, we recover precisely the value of $u(y)$ with this identity. In fact this procedure need not be restricted to bounded domains Ω. If we specify that $u \in C^2(\mathbb{R}^n)$ is such that $\lim_{R\to+\infty} \int_{S_R(0)} (\Gamma\partial_N u - \partial_N \Gamma\, u)\, dS_x = 0$ for $y \in \mathbb{R}^n$, then the statement of Theorem 4.3 can be interpreted distributionally in terms of the Dirac δ-function:

$$\Delta_x \Gamma(|x-y|) = \delta_y(x) .$$

Formula (4.12) has further implications having to do with the smoothness of solutions of Laplace's equation away from the boundary of their domain of definition, a topic known as the *interior regularity* of elliptic equations. Away from its pole at $x = y$, the fundamental solution Γ is C^∞ smooth; in fact it is C^ω, which is the notation for the class of real analytic functions. The results of Theorem 4.3 can be interpreted to deduce that a harmonic function is C^∞, indeed even real analytic, in the interior of its domain of definition; this is the conclusion of the following theorem.

Theorem 4.4. *Let $u \in C^2(\Omega) \cap C^1(\partial\Omega)$ be a harmonic function in Ω. Then in fact $u \in C^\omega$.*

Proof. For harmonic functions $u(x)$, identity (4.12) reads

$$u(y) = \int_{\partial\Omega} \big(u\, \partial_N \Gamma - \partial_N u\, \Gamma\big)\, dS_x .$$

Since $y \in \Omega$ in this expression, while the boundary integral only involves $x \in \partial\Omega$, then $\mathrm{dist}(x, y)$ is bounded from below and the RHS is clearly a C^ω function of y, hence the result. □

We may add any harmonic function $w(x)$ to the fundamental solution $\Gamma(|x-y|)$ and retain the above properties, as

$$\Delta_x(\Gamma + w) = \Delta_x \Gamma + \Delta_x w = \delta_y(x) .$$

In particular consider any ball $B_\rho(y) \subseteq \Omega$ within the domain of definition of a harmonic function $u(x)$. Use the quantity $\Gamma(|x-y|) - \Gamma(\rho)$ as a fundamental solution in (4.12) to find that

$$u(y) = \int_{S_\rho(y)} \big(\Gamma(|x-y|) - \Gamma(\rho)\big)\partial_N u \, dS_x - \int_{S_\rho(y)} u \, \partial_r \Gamma\Big|_{r=\rho} dS_x$$
$$= \int_{S_\rho(y)} u(x) \frac{1}{\omega_n \rho^{n-1}} \, dS_x \ .$$

This proves the following statement.

Theorem 4.5 (Gauss' law of arithmetic mean). *A harmonic function satisfies the integral identity*

$$u(y) = \frac{1}{\omega_n \rho^{n-1}} \int_{S_\rho(y)} u(x) \, dS_x = \frac{1}{\omega_n} \int_{v \in S_1(0)} u(y + \rho v) \, dS_v \tag{4.13}$$

for all $B_\rho(y)$ contained in the domain of definition of u. An alternative version of this statement is that

$$u(y) = \frac{n}{\omega_n \rho^n} \int_{B_\rho(y)} u(x) \, dx \ . \tag{4.14}$$

In words, this result states that a harmonic function $u(y)$ is equal to its average values over spheres $S_\rho(y)$ about y. This fact is consistent with discretizations of the Laplace operator in numerical simulations. For example, the simplest version of a finite difference approximation of the Laplacian at $y \in \mathbb{Z}^n$ is

$$\Delta_h u(y) := \frac{1}{h^2}\Big(\sum_{x \in \mathbb{Z}^n : |x-y|=1} u(x) - 2nu(y) \Big) \ .$$

A discrete harmonic function $\Delta_h u = 0$ explicitly satisfies the property that $u(y)$ is equal to the average of its values at the nearest neighbor points $|x - y| = 1$.

4.4. Maximum principle

The maximum principle is a recurring theme in the theory of elliptic and parabolic PDEs. It provides a strong and ubiquitous method to assure the uniqueness of solutions and to provide estimates on them and their derivatives.

Theorem 4.6 (Weak maximum principle). *Suppose that Ω is a bounded domain and that $u(x) \in C^2(\Omega) \cap C^0(\overline{\Omega})$ satisfies*

$$\Delta u \geq 0 \tag{4.15}$$

for $x \in \Omega$. Then the maximum value of $u(x)$ over the set $\overline{\Omega}$ is achieved for some $x \in \partial\Omega$; namely, for all $x \in \Omega$,

$$u(x) \leq \max_{x \in \partial\Omega} u(x) := M \ . \tag{4.16}$$

If $u(x) \in C^2(\omega) \cap C^0(\overline{\Omega})$ is harmonic, then both u and $-u$ satisfy the hypotheses of the theorem, and therefore there are upper and lower bounds on $u(x)$ that only depend upon its boundary values; namely, for $x \in \Omega$,

$$\min_{x\in\partial\Omega} u(x) \le u(x) \le \max_{x\in\partial\Omega} u(x) \ .$$

Proof. First of all, start with the stronger assumption of strict inequality, $\Delta u(x) > 0$ in Ω. Suppose that at some point $x_0 \in \Omega$ the function $u(x)$ achieves its maximum, $u(x_0) = \max_{x\in\overline{\Omega}}(u(x))$. Then $\nabla u(x_0) = 0$ and the Hessian matrix of u satisfies

$$H(u) = \big(\partial_{x_j}\partial_{x_\ell} u(x_0)\big)_{j\ell=1}^n \le 0 \ .$$

However, this contradicts the fact that $\Delta u > 0$; indeed,

$$\Delta u(x) = \operatorname{tr}\big(H(u)\big) > 0 \ .$$

Hence no such maximum point can exist in Ω, and this argument even rules out local maxima. For the general case where we assume only that $\Delta u \ge 0$ in Ω, consider the function $v(x) = u(x) + \varepsilon|x|^2$, which satisfies $\Delta v(x) = \Delta u(x) + 2\varepsilon n > 0$. Therefore v is covered by the hypotheses of the first case, from which we conclude

$$\begin{aligned}\max_{x\in\overline{\Omega}}(u(x)) &\le \max_{x\in\overline{\Omega}}(u(x)+\varepsilon|x|^2) \\ &= \max_{x\in\partial\Omega}(u(x)+\varepsilon|x|^2) \le \max_{x\in\partial\Omega}(u(x)) + \varepsilon \max_{x\in\partial\Omega}(|x|^2) \ .\end{aligned}$$

Since Ω is bounded, $\max_{x\in\partial\Omega}(|x|^2)$ is finite, and since ε is arbitrary, we have shown that

$$\max_{x\in\overline{\Omega}}(u(x)) \le \max_{x\in\partial\Omega}(u(x)) \ .$$

□

The advantage of this proof is that it immediately generalizes to many other elliptic equations, including

$$\sum_{j,\ell=1}^n a^{j\ell}(x)\partial_{x_j}\partial_{x_\ell}u + \sum_{j=1}^n b^j(x)\partial_{x_j}u = 0 \ ,$$

where the matrices $(a^{j\ell}(x))_{j,\ell=1}^n$ are positive definite. The disadvantage is that it relies upon the hypothesis that $u(x) \in C^2(\Omega)$, which in the present situation is stronger than necessary. The result on Gauss' law of arithmetic mean (4.13) in the previous section allows us to discuss maximum principles for functions which are not $C^2(\Omega)$ nor even $C^1(\Omega)$, but which merely are continuous.

Definition 4.7. A function $u \in C(\Omega)$ is *subharmonic* in Ω if for all $y \in \Omega$ and all $B_\rho(y) \subseteq \Omega$, one has

$$u(y) \leq \frac{1}{\omega_n \rho^{n-1}} \int_{S_\rho(y)} u(x)\, dS_x \ .$$

A function $u \in C(\Omega)$ is *superharmonic* in Ω if for all $y \in \Omega$ and all $B_\rho(y) \subseteq \Omega$, the opposite inequality holds:

$$u(y) \geq \frac{1}{\omega_n \rho^{n-1}} \int_{S_\rho(y)} u(x)\, dS_x \ .$$

This nomenclature is motivated by the fact that the graph of a subharmonic function lies beneath the graph of the harmonic function with the same boundary values on $\partial\Omega$, as we will show below. Similarly, superharmonic functions lie above harmonic functions sharing their boundary values. Furthermore, if we also knew that $u \in C^2(\Omega)$, then the inequality

$$\Delta u(x) \geq 0 \ , \qquad \big(\text{respectively, } \Delta u(x) \leq 0\big) \ ,$$

implies that $u(x)$ is subharmonic (respectively, superharmonic).

It turns out that subharmonic functions $u \in C(\Omega)$ satisfy the maximum principle, a statement that substantially lessens the necessary hypotheses for maximum principles and proves to be a very useful principle in nonlinear problems and in other generalizations of Laplace's equation.

Theorem 4.8 (Maximum principles for subharmonic functions). *Let Ω be a bounded domain and suppose that $u \in C(\overline{\Omega})$ is subharmonic. Then*

$$\max_{x\in\overline{\Omega}}(u(x)) = \max_{x\in\partial\Omega}(u(x)) \ . \tag{4.17}$$

Furthermore, if Ω is connected, then either for all $x \in \Omega$ we have

$$u(x) < \max_{x\in\partial\Omega}(u(x)) \ , \tag{4.18}$$

or else, if at some point $x \in \Omega$ equality holds, then $u(x)$ is necessarily a constant function $u(x) \equiv M := \max_{x\in\partial\Omega}(u(x))$.

The latter result is known as the strong maximum principle for subharmonic function. If $u(x)$ is superharmonic, then $-u(x)$ is subharmonic, and therefore u satisfies a *minimum principle.* A corollary of Theorem 4.8 is that harmonic functions, which are both sub- and superharmonic, satisfy upper and lower estimates in the supremum norm. Namely, if $u(x)$ is harmonic in Ω and $u(x) = f(x)$ for $x \in \partial\Omega$, then the weak maximum principle implies that

$$\min_{x\in\partial\Omega}(f(x)) \leq u(x) \leq \max_{x\in\partial\Omega}(f(x))$$

for all $x \in \overline{\Omega}$.

Corollary 4.9 (Strong maximum principle). *Let Ω be a bounded connected domain, and suppose that $u \in C(\Omega)$ is harmonic and satisfies $u(x) = f(x)$ for $x \in \partial\Omega$. That is,*

$$\Delta u = 0 \quad \text{for} \quad x \in \Omega \ .$$

Then for all $x \in \Omega$,

$$\min_{x\in\partial\Omega}(f(x)) < u(x) < \max_{x\in\partial\Omega}(f(x)) \ .$$

Or else, if equality holds at any point $x \in \Omega$ instead of either of these strict inequalities, then

$$u(x) \equiv M := \min_{x\in\partial\Omega}(f(x)) = \max_{x\in\partial\Omega}(f(x)) \ .$$

Proof of Theorem 4.8. The proof of this second version of the maximum principle is more specific to Laplace's equation, using the property of subharmonicity. Consider Ω a domain which is connected and $u(x)$ a subharmonic function, and set $M := \sup_{x\in\Omega}(u(x))$. Decompose the domain into two disjoint subsets:

$$\Omega = \{x \in \Omega \,:\, u(x) = M\} \cup \{x \in \Omega \,:\, u(x) < M\} := \Omega_1 \cup \Omega_2 \ .$$

Since $u(x) \in C(\Omega)$, then Ω_2 is open as a subset of Ω because an inequality is an open condition. The claim is that the set Ω_1 is also open in Ω, which we will prove using the property of subharmonicity. Therefore, because of the fact of being connected, either $\Omega = \Omega_2$ and $\Omega_1 = \emptyset$, whereupon strict inequality holds throughout Ω, or else $\Omega = \Omega_1$ and $\Omega_2 = \emptyset$, in which case $u(x) = M$ a constant.

To prove the claim above that Ω_1 is open, consider $x_0 \in \Omega_1$ and refer to the property of subharmonicity on all sufficiently small spheres $S_\rho(x_0)$ about x_0 which lie in Ω. This property implies that

$$0 \le \frac{1}{\omega_n \rho^{n-1}} \int_{|x-x_0|=\rho} u(x) - u(x_0)\, dS_x \ .$$

The integrand is nonpositive because $u(x) \le M = u(x_0)$ in Ω. But the integrand cannot be negative anywhere near x_0 either, for that would violate the above inequality. Therefore we must have $S_\rho(x_0) \subseteq \Omega_1$ for all sufficiently small ρ, which is the statement that Ω_1 is open. □

4.5. Green's functions and Dirichlet–Neumann operators

As we observed in section 4.3, we may add any harmonic function to the fundamental solution $\Gamma(|x-y|)$ and retain the property that $\Delta(\Gamma + w) = \delta_y$. This is useful because the fundamental solution $\Gamma(|x-y|)$ doesn't satisfy the proper Dirichlet boundary conditions on domains Ω, with the only exception being the case $\Omega = \mathbb{R}^n$. Thus we may add a function $w(x,y) \in C^2(\Omega \times \Omega)$ that satisfies

$$\begin{aligned} \Delta_x w(x,y) &= 0 \quad \text{for} \quad x \in \Omega \ , \\ w(x,y) &= -\Gamma(|x-y|) \quad \text{for} \quad x \in \partial\Omega \ . \end{aligned}$$

The function $w(x,y)$ is harmonic in $x \in \Omega$ and depends parametrically on $y \in \Omega$. Then

$$G(x,y) := \Gamma(|x-y|) + w(x,y)$$

is still a fundamental solution in the distributional sense as in (4.12), meaning that $\Delta_x G(x,y) = \delta_y(x)$. Furthermore $G(x,y) = 0$ for $x \in \partial\Omega$, and therefore

$$u(x) := \int_\Omega G(x,y)h(y)\,dy$$

satisfies

$$\begin{aligned} \Delta_x u(x) &= \Delta_x \int_\Omega G(x,y)h(y)\,dy \\ &= \Delta_x \int_\Omega \Gamma(|x-y|)h(y)\,dy + \int_\Omega \Delta_x w(x,y)h(y)\,dy = h(x) \ . \end{aligned}$$

It also satisfies Dirichlet boundary conditions; indeed, for $x \in \partial\Omega$,

$$u(x) = \int_\Omega G(x,y)h(y)\,dy = 0 \ .$$

The function $G(x,y)$ is called the *Green's function* for the domain Ω; it is the integral kernel of the solution operator $\mathbf{P}$ for the Poisson problem (4.2),

$$u(x) = \int_\Omega G(x,y)h(y)\,dy := \mathbf{P}h(x) \ . \tag{4.19}$$

The Green's function for $\mathbb{R}^n_+$. While it is usually not possible to have explicit formulae for the Green's function $G(x,y)$ for a general domain, it is straightforward for the domain consisting of the half space $\mathbb{R}^n_+$. Let $y = (y', y_n) \in \mathbb{R}^n_+$ so that $y_n > 0$, and denote the point that is its reflection through the boundary $\{x_n = 0\}$ by $y^* := (y', -y_n) \in \mathbb{R}^n_-$. The function $w(x,y) = -\Gamma(|x-y^*|)$ satisfies the property that it is harmonic in $\mathbb{R}^n_+$ (since its singularity at $x = y^*$ is in $\mathbb{R}^n_-$), and $\Gamma(|x-y^*|) = \Gamma(|x-y|)$ when $x = (x', 0) \in \partial\mathbb{R}^n_+$. This gives rise to the following expression for the Green's function for the domain $\mathbb{R}^n_+$:

$$\begin{aligned} G(x,y) &= \Gamma(|x-y|) - \Gamma(|x-y^*|) \\ &= \frac{1}{(2-n)\omega_n}\Big(\frac{1}{(|x'-y'|^2 + (x_n - y_n)^2)^{(n-2)/2}} \\ &\qquad - \frac{1}{(|x'-y'|^2 + (x_n + y_n)^2)^{(n-2)/2}}\Big) \end{aligned} \tag{4.20}$$

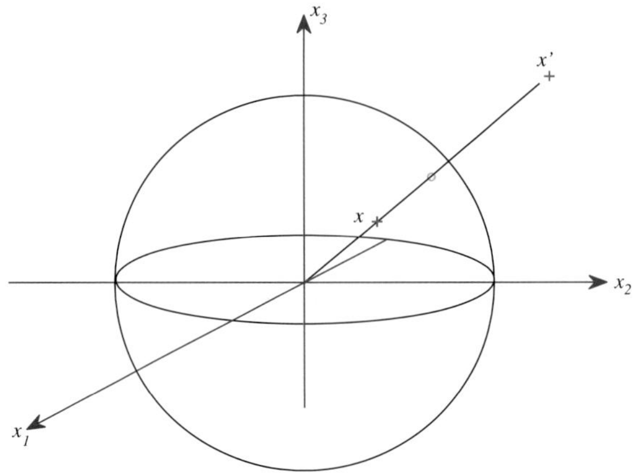

Figure 2. Conformal inversion of the point $x \in B_R(0)$ whose image is x^*.

when $n \geq 3$. When $n = 2$,

$$G(x,y) = \frac{1}{4\pi}\log\Big(\frac{(x_1-y_1)^2+(x_2-y_2)^2}{(x_1-y_1)^2+(x_2+y_2)^2}\Big) . \tag{4.21}$$

This procedure is similar to the method of images, as one can see. Actually, in two dimensions there is a connection with the theory of complex variables; because of the Riemann mapping theorem there is an expression for the Green's function for arbitrary simply connected domains in terms of formula (4.21) and the Riemann mapping of the domain to the upper half plane $\mathbb{R}^2_+$.

The Green's function for $B_R(0)$. A second case of a domain Ω for which the Green's function has an explicit expression is the ball $B_R(0) \subseteq \mathbb{R}^n$. Given $x, y \in \Omega$ and another point $y^* \notin \Omega$, the function constructed from the fundamental solution $\Gamma(|x-y|)$ given by

$$G(x,y) = \Gamma(|x-y|) - c(y)\Gamma(|x-y^*|)$$

is harmonic as a function of $x \in \Omega$,

$$\Delta_x G(x,y) = 0 \ , \qquad x \neq y \ ,$$

with the singularity of the fundamental solution at $x = y$. In the special case of $\Omega = B_R(0)$, one may choose the exterior point y^* to be the inversion of y in the sphere $\mathbb{S}^{n-1}_R = \partial B_R(0)$, given by

$$y^* = \frac{R^2}{|y|^2}y \ .$$

For $x \in \mathbb{S}^{n-1}_R$, this choice satisfies the special geometrical property that

$$\frac{|x-y^*|}{|x-y|} = \frac{R}{|y|} \ , \tag{4.22}$$

the RHS being an expression independent of x. This is illustrated in Figure 2. Therefore if we take $c(y) = \left(\frac{R}{|y|}\right)^{n-2}$, then for $x \in \mathbb{S}_R^{n-1}$ we will have

$$c(y)\Gamma(|x - y^*|) = c(y)\frac{1}{(2-n)\omega_n}\frac{1}{|x - y^*|^{n-2}}$$
$$= c(y)\frac{1}{(2-n)\omega_n}\frac{1}{|x-y|^{n-2}}\left(\frac{|y|}{R}\right)^{n-2} = \frac{1}{(2-n)\omega_n}\frac{1}{|x-y|^{n-2}} ,$$

and therefore $G(x, y)$ will have the properties that it is harmonic in the x-variable and that $G(x, y) = 0$ for $x \in \mathbb{S}_R^{n-1}$. This gives the expression for the Green's function for the ball $B_R(0)$,

$$G(x, y) = \Gamma(|x - y|) - c(y)\Gamma(|x - y^*|)$$
$$= \frac{1}{(2-n)\omega_n}\left(\frac{1}{|x - y|^{n-2}} - \frac{R^{n-2}}{|y|^{n-2}}\frac{1}{|x - y^*|^{n-2}}\right) . \tag{4.23}$$

Although it is not so evident for the Green's function for an arbitrary domain Ω, or from the way we constructed the Green's function for special domains Ω, the Green's function has the symmetry that $G(x, y) = G(y, x)$. This reflects the property of the Laplace operator with Dirichlet boundary conditions being a self-adjoint operator.[1] This property can be verified explicitly for the Green's functions for the upper half space and for $B_R(0)$ upon inspecting the formulae in (4.20), (4.21), and (4.23).

The Poisson kernel. The Green's function $G(x, y)$ is the integral kernel for the solution of the Poisson problem (4.2), but it also is relevant for the Dirichlet problem (4.1). Let $u \in C^2(\Omega) \cap C^1(\overline{\Omega})$ be harmonic. Then by Green's identities

$$0 = \int_\Omega G(x, y)\Delta u(x)\, dx$$
$$= u(y) + \int_{\partial\Omega} \left(G(x, y)\partial_N u(x) - \partial_{N_x} G(x, y)u(x)\right) dS_x .$$

Using the fact that $G(x, y) = 0$ for $x \in \partial\Omega$, we deduce that

$$u(y) = \int_{\partial\Omega} u(x)\partial_{N_x} G(x, y)\, dS_x$$
$$= \int_{\partial\Omega} f(x)\partial_{N_x} G(x, y)\, dS_x := \mathbf{D}f(y) , \tag{4.24}$$

where $\mathbf{D}f(y)$ is the operator giving the solution to the Dirichlet problem (4.1) for $f(x)$ Dirichlet data for $u(x)$ on the boundary $\partial\Omega$. The quantity $D(x, y) := \partial_{N_x} G(x, y)$ in the integrand is called the *Poisson kernel* for the domain Ω.

[1]Technically, the Laplace operator is a symmetric operator which is self-adjoint when restricted to an appropriate subspace $H_0^1(\Omega) \subseteq L^2(\Omega)$.

In the case that $\Omega = \mathbb{R}^n_+$, the explicit form of the Green's function (4.20) (and (4.21) in the case in which $n = 2$) gives an equally explicit expression for the Poisson kernel. This is

$$-\partial_{x_n} G(x,y)\Big|_{x_n=0} = D(x'-y', y_n) = \frac{2}{\omega_n}\left(\frac{y_n}{(|x'-y'|^2 + y_n^2)^{n/2}}\right) . \tag{4.25}$$

In the case of the ball $\Omega = B_R(0)$, the Poisson kernel in (4.24) is given by the expression

$$D(x,y) = \frac{1}{R\omega_n}\left[\frac{R^2 - |y|^2}{|x-y|^n}\right] , \tag{4.26}$$

as one finds from (4.22) and (4.23) after a short calculation.

The Neumann problem. There is a related expression for the analog of the Green's function for the Neumann problem, which can be found by choosing another function $w(x,y)$ which is harmonic in x and forming the fundamental solution

$$N(x,y) = \Gamma(|x-y|) + w(x,y) .$$

If $w(x,y)$ is chosen so that $\partial_N \Gamma(|x-y|) = -\partial_N w(x,y)$ for all $x \in \partial\Omega$, then the function $N(x,y)$ is the integral kernel for the solution operator of the Neumann problem (4.3). That is, the analog of the formula (4.24) holds. On general domains, given Neumann data $g(x)$ such that $\int_{\partial\Omega} g(x)\, dS_x = 0$,

$$\begin{aligned} u(y) &= -\int_{\partial\Omega} N(x,y)\partial_N u(x)\, dS_x \\ &= -\int_{\partial\Omega} N(x,y) g(x)\, dS_x := \mathbf{N}g(y) , \end{aligned} \tag{4.27}$$

analogous to (4.24), as one shows with a calculation using Green's second identity. In the case that $\Omega = \mathbb{R}^n_+$, the choice is that $w(x,y) = \Gamma(|x-y^*|)$, and

$$\begin{aligned} N(x,y) &= \Gamma(|x-y|) + \Gamma(|x-y^*|) \\ &= \frac{1}{(2-n)\omega_n}\Bigg(\frac{1}{(|x'-y'|^2 + (x_n - y_n)^2)^{(n-2)/2}} \\ &\qquad + \frac{1}{(|x'-y'|^2 + (x_n + y_n)^2)^{(n-2)/2}}\Bigg) \end{aligned} \tag{4.28}$$

when $n \geq 3$. When $n = 2$,

(4.29)
$$N(x,y) = \frac{1}{4\pi}\log\Big(\big((x_1-y_1)^2 + (x_2-y_2)^2\big)\big((x_1-y_1)^2 + (x_2+y_2)^2\big)\Big) .$$

Setting $x_n = 0$ gives rise to the *single layer potential*

$$S(x'-y', y_n) = \frac{2}{(2-n)\omega_n}\left(\frac{1}{(|x'-y'|^2 + y_N^2)^{(n-2)/2}}\right) \tag{4.30}$$

with which one solves the Neumann boundary value problem with data $g(x)$, namely

$$u(y) = \int_{x' \in \mathbb{R}^{n-1}} S(x' - y', x_n) g(x')\, dx' := \mathbf{S}g(y) \ .$$

The physical interpretation of this expression is that it represents the electrostatic potential induced by a charge distribution of density $g(x')$ on the boundary $\partial\mathbb{R}^n_+ = \mathbb{R}^{n-1}$ of the domain $\mathbb{R}^n_+$. Similarly, the *double layer potential* is given by

$$u(y) = \int_{x' \in \mathbb{R}^{n-1}} \frac{1}{\omega_n} \Big(\frac{y_n}{(|x' - y'|^2 + y_n^2)^{n/2}} \Big) f(x')\, dx' = \frac{1}{2} \mathbf{D} f(y) \ ,$$

and its interpretation is the resulting electrostatic potential from a distribution of dipole charges with density $f(x')$ on the boundary of $\mathbb{R}^n_+$.[2] The potential associated to a distribution $\rho(x)$ of charge in $\mathbb{R}^n$ is given by a convolution with the fundamental solution itself

$$u(y) = \int_{\mathbb{R}^n} \Gamma(x - y) \rho(x)\, dx \ .$$

The Dirichlet–Neumann operator. Given Dirichlet data on the boundary of a domain Ω, it is often the most important part of the solution process for $u(x)$ of the Dirichlet problem to recover the normal derivatives of the solution $\partial_N u(x)$ on the boundary. For Ω a conducting body, this would be the map from applied voltage $u(x) = f(x)$ on $\partial\Omega$ to the resulting current $\partial_N u(x) = g(x)$ across the boundary. This map can be expressed in terms of the Green's function; in particular, using the Poisson kernel we can express the solution to the Dirichlet problem,

$$u(y) = \int_{\partial\Omega} (N_x \cdot \nabla_x) G(x, y) f(x)\, dS_x \ .$$

Therefore we have an expression for its normal derivative on $\partial\Omega$, namely
(4.31)

$$(N_y \cdot \nabla_y) u(y) \Big|_{y \in \partial\Omega} = \int_{\partial\Omega} (N_x \cdot \nabla_x)(N_y \cdot \nabla_y) G(x, y) f(x)\, dS_x := \mathbf{G} f(y) \ ,$$

[2] In more generality, the single layer potential on a codimension-one hypersurface Λ is given by

$$u(x) = \int_\Lambda \frac{1}{|x - y|^{n-2}} g(y)\, dS_y \ ,$$

corresponding to the potential $u(x)$ due to a charge distribution of density $f(x)$ on Λ. The double layer potential is given by

$$u(x) = \int_\Lambda g(y)\, N \cdot \partial_y \frac{1}{|x - y|^{n-2}}\, dS_y \ ,$$

which is interpreted physically as the potential due to a dipole distribution along Λ with density $g(x)$. In the case that $\Lambda = \mathbb{R}^{n-1} \subseteq \mathbb{R}^n$, these two coincide with the kernel for the Neumann problem and the Poisson kernel, respectively. However, this relationship does not hold in general.

where $\mathbf{G}$ is the Dirichlet–Neumann operator for the domain Ω. There is symmetry in the exchange of x with y in the integrand of (4.31), from which we deduce that the Dirichlet–Neumann operator is self-adjoint;[3] $\mathbf{G}^T = \mathbf{G}$.

Recall the energy identity for a harmonic function $u(x)$,

$$(4.32)\quad 0 \le \int_\Omega |\nabla u(x)|^2\,dx = \int_{\partial\Omega} u(x)\partial_N u(x)\,dS_x = \int_{\partial\Omega} f(x)(\mathbf{G}f)(x)\,dS_x\ .$$

This is to say that the operator $\mathbf{G}$ is nonnegative definite as well as self-adjoint. Formula (4.32) also exhibits the relation between the Dirichlet integral of a harmonic function over a domain Ω and the boundary integral over $\partial\Omega$ involving the Dirichlet–Neumann operator.

4.6. Poisson kernel on $\mathbb{R}^n_+$

This section is dedicated to techniques based on the Fourier transform. For Laplace's equation we can do this directly only in particular cases, the most straightforward being that the domain $\Omega = \mathbb{R}^n_+ := \{x = (x_1, \dots, x_n) \in \mathbb{R}^n : x_n > 0\}$, the half space, and this is the situation that we will discuss. It would also be possible to directly use similar techniques based on Fourier series to solve Laplace's equation on the disk $\mathbb{D}^2 := \{x \in \mathbb{R}^2 : |x| < 1\}$ or the polydisc $\mathbb{D}^{2n} := \{z = (x_1, \dots, x_n, y_1, \dots, y_n) \in \mathbb{R}^{2n} : (x_j^2 + y_j^2) < 1\ , j = 1, \dots, n\}$. We will, however, stick with the half space $\mathbb{R}^n_+$.

The Dirichlet problem (4.1) on $\mathbb{R}^n_+$ is to solve

$$(4.33)\qquad \begin{aligned} &\Delta u(x) = 0\ , && x = (x_1, \dots, x_n) \in \mathbb{R}^n_+\ ,\\ &u(x) = f(x')\ , && x = (x', 0)\ ,\quad x' \in \mathbb{R}^{n-1} = \partial\mathbb{R}^n_+\ . \end{aligned}$$

We will assume that $u(x)$ and $\nabla u(x)$ tend to zero as $x_n \to +\infty$. For the special boundary data $f(x') = e^{i\xi'\cdot x'}$ with $\xi' \in \mathbb{R}^{n-1}$, there are explicit solutions

$$(4.34)\qquad u(x', x_n) = e^{i\xi'\cdot x'} e^{-|\xi'|x_n}$$

since

$$\begin{aligned} \Delta u(x', x_n) &= \sum_{j=1}^{n-1} \partial^2_{x_j}\left(e^{i\xi'\cdot x'} e^{-|\xi'|x_n}\right) + \partial^2_{x_n}\left(e^{i\xi'\cdot x'} e^{-|\xi'|x_n}\right)\\ &= \Big(\sum_{j=1}^{n-1} -\xi_j'^2 + |\xi'|^2\Big)\left(e^{i\xi'\cdot x'} e^{-|\xi'|x_n}\right) = 0\ . \end{aligned}$$

The other possible solution is $e^{i\xi'\cdot x'} e^{+|\xi'|x_n}$, but this is ruled out by its growth as $x_n \to +\infty$. The Fourier transform allows us to decompose a general

[3] Analogous to the case above, technically the operator $\mathbf{G}$ is symmetric on $C^1(\partial\Omega)$ and self-adjoint on an appropriate subspace $H^{1/2}(\partial\Omega) \subseteq L^2(\partial\Omega)$.

function $f(x')$ on the boundary (in $L^2(\mathbb{R}^{n-1})$ or perhaps in $L^1(\mathbb{R}^{n-1})$) into a composite of complex exponentials

$$f(x') = \frac{1}{\sqrt{2\pi}^{n-1}} \int e^{i\xi'\cdot x'} \hat{f}(\xi')\, d\xi' \, .$$

By using (4.34), the general solution $u(x)$ can be expressed as a superposition of the above special solutions with boundary data expressed via the Fourier transform, namely

$$\begin{aligned} u(x) &= \frac{1}{\sqrt{2\pi}^{n-1}} \int e^{i\xi'\cdot x'} e^{-|\xi'|x_n} \hat{f}(\xi')\, d\xi' \\ &= \frac{1}{(2\pi)^{n-1}} \int \Big(\int e^{i\xi'\cdot(x'-y')} e^{-|\xi'|x_n}\, d\xi' \Big) f(y')\, dy' \\ &= \int D(x'-y', x_n) f(y')\, dy' = \mathbf{D}f(x) \, . \end{aligned} \tag{4.35}$$

The function $D(x', x_n)$ is an integral kernel for the solution operator, often called the Poisson kernel for $\mathbb{R}^n_+$ or the double layer potential, and the solution $u(x)$ is evidently given by convolution of the Dirichlet boundary data with $D(x', x_n)$. Evaluating the above Fourier integral expression (4.35), we get an explicit expression:

$$\begin{aligned} D(x', x_n) &= \frac{1}{(2\pi)^{n-1}} \int e^{i\xi'\cdot x'} e^{-|\xi'|x_n}\, d\xi' \\ &= \frac{2}{\omega_n} \Big(\frac{x_n}{(|x'|^2 + x_n^2)^{n/2}} \Big) \, . \end{aligned} \tag{4.36}$$

The second line of (4.36) has already been derived as (4.25) in section 4.5 using the method of images. The constant ω_n is the surface area of the unit sphere in $\mathbb{R}^n$.

Theorem 4.10. *The solution of the Dirichlet problem on $\mathbb{R}^n_+$ with data $f(x') \in L^2(\mathbb{R}^{n-1})$ is given by*

$$\begin{aligned} u(x) &= \int_{\mathbb{R}^{n-1}} D(x'-y, x_n) f(y')\, dy' \\ &= \frac{1}{(2\pi)^{n-1}} \int_{\mathbb{R}^{n-1}} \Big(\int_{\mathbb{R}^{n-1}} e^{i\xi'\cdot(x'-y')} e^{-|\xi'|x_n}\, d\xi' \Big) f(y')\, dy' \, . \end{aligned} \tag{4.37}$$

For $x_n > 0$, this function is C^∞ (differentiable an arbitrary number of times).

Proof. For $x_n > 0$, both of the above integrals in (4.37) converge absolutely, as indeed we have

$$\int_{\mathbb{R}^{n-1}} \big| e^{i\xi'\cdot x'} e^{-|\xi'|x_n} \hat{f}(\xi') \big|\, d\xi' \le \|\hat{f}\|_{L^2(\mathbb{R}^{n-1})} \Big(\int_{\mathbb{R}^{n-1}} e^{-2|\xi'|x_n}\, d\xi' \Big)^{1/2}$$

(where we have used the Cauchy–Schwarz inequality and the Plancherel identity (3.16)), and hence we also learn that the solution $u(x)$ has an upper bound in $\mathbb{R}^n_+$ which quantifies its decay rate as $x_n \to +\infty$. Namely,

$$|u(x', x_n)| \le \Big(\int_{\mathbb{R}^{n-1}} e^{-2|\xi'|x_n}\, d\xi'\Big)^{1/2} \|\hat{f}\|_{L^2(\mathbb{R}^{n-1})} \le \frac{C}{|x_n|^{(n-1)/2}} \|f\|_{L^2(\mathbb{R}^{n-1})} \ .$$

Further derivatives of expression (4.37) do not change the properties of absolute convergence of the integral for $x_n > 0$, and we verify that the function we have produced is indeed harmonic:

$$\Delta u(x) = \frac{1}{\sqrt{2\pi}^{n-1}} \int_{\mathbb{R}^{n-1}} \Delta\big(e^{i\xi'\cdot x'} e^{-|\xi'|x_n}\big) \hat{f}(\xi')\, d\xi' = 0 \ .$$

The issue is whether the harmonic function $u(x', x_n)$ that we have produced converges to $f(x')$ as $x_n \to 0$, and in what sense. In this paragraph we will show that for every $x_n > 0$, $u(x', x_n)$ is an $L^2(\mathbb{R}^{n-1})$ function of the horizontal variables $x' \in \mathbb{R}^{n-1}$, and that $u(x', x_n) \to f(x')$ in the $L^2(\mathbb{R}^{n-1})$ sense as $x_n \to 0$. By Plancherel,

$$\|u(x', x_n) - f(x')\|_{L^2(\mathbb{R}^{n-1})} = \|e^{-|\xi'|x_n} \hat{f}(\xi') - \hat{f}(\xi')\|_{L^2(\mathbb{R}^{n-1})} \ .$$

Because $\|\hat{f}\|_{L^2} < +\infty$, for any $\delta > 0$ there is a (possibly large) $R > 0$ such that

$$\int_{|\xi'|>R} |\hat{f}(\xi')|^2\, d\xi' < \delta \ .$$

We now estimate

$$\begin{aligned}
&\|u(x', x_n) - f(x')\|^2_{L^2(\mathbb{R}^{n-1})} = \|\big(e^{-|\xi'|x_n} - 1\big)\hat{f}(\xi')\|^2_{L^2(\mathbb{R}^{n-1})} \\
&= \int_{|\xi'|\le R} \big|\big(e^{-|\xi'|x_n} - 1\big)\hat{f}(\xi')\big|^2\, d\xi' + \int_{|\xi'|>R} \big|\big(e^{-|\xi'|x_n} - 1\big)\hat{f}(\xi')\big|^2\, d\xi' \\
&\le \big|e^{-Rx_n} - 1\big| \int_{|\xi'|\le R} |\hat{f}(\xi')|^2\, d\xi' + 2\delta \ .
\end{aligned}$$

The first term of the RHS vanishes in the limit as $\delta \to 0$. Since δ is arbitrary, we conclude that $u(x', x_n)$ converges to $f(x')$ as $x_n \to 0$ in the $L^2(\mathbb{R}^{n-1})$ sense, which is that the L^2-norm of their difference tends to zero. □

Dirichlet–Neumann operator. It is useful to work out in detail the form of the Dirichlet–Neumann operator on the domain $\mathbb{R}^n_+$. The solution to the Dirichlet problem for $\mathbb{R}^n_+$ is given in (4.37) in terms of the Fourier transform of the Poisson kernel:

$$u(x) = \frac{1}{\sqrt{2\pi}^{n-1}} \int_{\mathbb{R}^{n-1}} e^{i\xi'\cdot x'} e^{-|\xi'|x_n} \hat{f}(\xi')\, d\xi' \ .$$

Recall that $\widehat{\frac{1}{i}\partial_{x_j} u}(\xi) = \xi_j \hat{u}(\xi)$, which motivates the notation for differential operators that $D_j = \frac{1}{i}\partial_{x_j}$ and the definition of a general Fourier multiplier operator

$$\begin{aligned}\big(m(D')f\big)(x') &= \frac{1}{\sqrt{2\pi}^{n-1}} \int_{\mathbb{R}^{n-1}} e^{i\xi'\cdot x'} m(\xi')\hat{f}(\xi')\, d\xi' \\ &= \frac{1}{(2\pi)^{n-1}} \iint_{\mathbb{R}^{n-1}\times\mathbb{R}^{n-1}} e^{i\xi'\cdot(x'-y')} m(\xi') f(y')\, d\xi' dy' \ .\end{aligned}$$

In these terms the harmonic extension $u(x)$ of the boundary conditions $f(x')$ can be written as

$$u(x) = e^{-x_n|D'|} f(x') := (\mathbf{D}f)(x', x_n) \ .$$

The operator $\mathbf{D}$ extends the Dirichlet data $f(x')$ to a harmonic function $u(x) = \mathbf{D}f(x', x_n)$ in the upper half space $\mathbb{R}^n_+$. This is an interpretation of formula (4.37) which uses a Fourier integral expression for the Poisson kernel.

Differentiating (4.37) with respect to x_n and evaluating the result on the boundary $\{x_n = 0\}$, the Dirichlet–Neumann operator has a related Fourier integral expression, namely

$$\begin{aligned}\mathbf{G}f(x') &= -\partial_{x_n}\big|_{x_n=0}\Big(\frac{1}{\sqrt{2\pi}^{n-1}} \int_{\mathbb{R}^{n-1}} e^{i\xi'\cdot x'} e^{-|\xi'|x_n} \hat{f}(\xi')\, d\xi'\Big) \\ &= \big(|D'|f\big)(x') \ .\end{aligned} \tag{4.38}$$

Finally, the Dirichlet integral can be expressed in terms of Fourier multipliers, using that

$$\begin{aligned}\partial_{x'} u(x', x_n) &= \frac{1}{\sqrt{2\pi}^{n-1}} \int_{\mathbb{R}^{n-1}} e^{i\xi'\cdot x'} i\xi' e^{-|\xi'|x_n} \hat{f}(\xi')\, d\xi' \ , \\ \partial_{x_n} u(x', x_n) &= \frac{1}{\sqrt{2\pi}^{n-1}} \int_{\mathbb{R}^{n-1}} e^{i\xi'\cdot x'} |\xi'| e^{-|\xi'|x_n} \hat{f}(\xi')\, d\xi' \ .\end{aligned}$$

Therefore

$$\begin{aligned}&\int_{\mathbb{R}^n_+} |\nabla u(x)|^2\, dx \\ &\quad = \int_0^{+\infty} \Big(\int_{\mathbb{R}^{n-1}} |i\xi'\hat{f}(\xi')|^2 e^{-2|\xi'|x_n} + |\xi'|^2 |\hat{f}(\xi')|^2 e^{-2|\xi'|x_n}\, d\xi'\Big) dx_n \ ,\end{aligned}$$

where we have used the Plancherel identity on the hyperplanes $\{(x', x_n) : x_n = \text{Constant}\}$. Thus

$$\int_{\mathbb{R}^n_+} |\nabla u(x)|^2 \, dx = \int_{\mathbb{R}^{n-1}} 2|\xi'|^2 |\hat{f}(\xi')|^2 \Big(\int_0^{+\infty} e^{-2|\xi'|x_n} \, dx_n \Big) d\xi$$
$$= \int_{\mathbb{R}^{n-1}} f(x') \big(|D'| f(x') \big) \, dx' \, .$$

4.7. Maximum principle again

A property that is evident of the Poisson kernel in (4.25) is that for $x_n > 0$,

$$D(x' - y', x_n) = \frac{2}{\omega_n} \Big(\frac{x_n}{(|x'|^2 + x_n^2)^{n/2}} \Big) > 0 \, .$$

Therefore whenever a solution of the Dirichlet problem (4.33) has the property that $f(x') \geq 0$,

$$u(x) = \int_{\mathbb{R}^{n-1}} D(x' - y', x_n) f(y') \, dy' > 0 \, ;$$

this follows from an argument that is very similar to the one we used for Theorem 3.5 on the heat equation.

Theorem 4.11. *Suppose that $f(x') \geq 0$; then for $x \in \mathbb{R}^n_+$ we have $u(x) \geq 0$, and in fact if at any point $x \in \mathbb{R}^n_+$ (meaning, with $x_n > 0$) it happens that $u(x) = 0$, then we conclude that $u(x) \equiv 0$ and $f(x') = 0$.*

From this result we have a comparison between solutions.

Corollary 4.12. *Suppose that $f_1(x') \leq f_2(x')$ for all $x' \in \mathbb{R}^{n-1}$. Then either*

$$u_1(x) = (D * f_1)(x) < u_2(x) = (D * f_2)(x)$$

on all of $\mathbb{R}^n_+$, or else $u_1(x) \equiv u_2(x)$ if equality holds at any point $x \in \mathbb{R}^n_+$. In particular if $f_1 = f_2$, then both $f_1 \leq f_2$ and $f_1 \geq f_2$ so that $u_1 \equiv u_2$.

This is our third encounter with the recurring theme in elliptic and parabolic PDEs of comparison and maximum principles. In the case of Laplace's equation and other elliptic equations, analog results closely hold on essentially arbitrary domains as well, although the Poisson kernel is not generally so explicit and the proof is different. In our present setting, the form of the Poisson kernel gives us a lower bound on the decay rates of solutions $u(x', x_n)$ for large x_n.

Corollary 4.13. *Suppose that $f(x') \geq 0$ and that $A \subseteq \{x' : f(x') \geq \delta\}$ is a bounded set of positive measure* $\operatorname{meas}(A) > 0$. *Then*

$$u(x', x_n) \geq \frac{2}{\omega_n} \Big(\frac{x_n}{\sup_{y' \in A} (|x' - y'|^2 + x_n^2)^{n/2}} \Big) \delta \operatorname{meas}(A) \, ,$$

and, in particular, solutions $u(x', x_n)$ that are positive have a minimum decay rate that is bounded below by $C|x_n|^{-n}$; that is, they cannot decay too rapidly as $x_n \to +\infty$.

Proof. Express the solution in terms of the Poisson kernel

$$\begin{aligned} u(x) &= \int_{\mathbb{R}^{n-1}} D(x' - y', x_n) f(y')\, dy' \\ &\geq \int_A D(x' - y', x_n) f(y')\, dy' \\ &\geq \inf_{y' \in A} \big(D(x' - y', x_n)\big) \int_A f(y')\, dy' \ , \end{aligned}$$

and of course

$$\int_A f(y')\, dy' \geq \delta \operatorname{meas}(A)$$

while

$$\inf_{y' \in A} \big(D(x' - y', x_n)\big) \geq \left(\frac{x_n}{\sup_{y' \in A}(|x' - y'|^2 + x_n^2)^{n/2}} \right) .$$

□

4.8. Oscillation and attenuation estimates

Another principle exhibited by solutions of Laplace's equation is the property of attenuation of oscillatory data. This is a feature that is related to the interior elliptic regularity of solutions of elliptic equations. It also is very relevant to applications such as imaging strategies in electrical impedence tomography, which is a medical imaging technique where the idea is to use electrostatic potentials to probe the interior of a patient's body in real time.

Theorem 4.14. *Suppose that $f(x') \in L^2(\mathbb{R}^{n-1})$ is the Dirichlet data for (4.33) on $\mathbb{R}^n_+$, and suppose in addition that*

$$\operatorname{dist}\big(\operatorname{supp}(\hat{f}(\xi')), 0\big) > \rho \ .$$

Then the solution $u(x', x_n)$ decays as $x_n \to +\infty$ with the upper bounds

$$|u(x', x_n)| \leq \frac{C}{|x_n|^{(n-1)/2}} e^{-\rho|x_n|} \ . \tag{4.39}$$

This estimate (4.39) gives an effective penetration depth of the solution $u(x)$ into the interior of the domain, in the situation in which the Dirichlet data $f(x')$ has no low frequency component, namely, that the Fourier transform $\hat{f}(\xi')$ vanishes in a neighborhood of $\xi' = 0$.

Proof. Since $\operatorname{dist}\big(\operatorname{supp}(\hat{f}(\xi')),0\big) > \rho$, there is a $\delta > 0$ such that $\inf_{\xi'\in\operatorname{supp}(\hat{f})}|\xi| > \rho(1+\delta)$. Using the Fourier representation for $u(x)$,

$$u(x) = \frac{1}{\sqrt{2\pi}^{n-1}} \int_{\operatorname{supp}(\hat{f})} e^{i\xi'\cdot x'} e^{-|\xi'|x_n} \hat{f}(\xi')\, d\xi' ,$$

we conclude that

$$\begin{aligned}
|u(x',x_n)| &\le \frac{1}{\sqrt{2\pi}^{n-1}} \int_{\operatorname{supp}(\hat{f})} \left|e^{-|\xi'|x_n} \hat{f}(\xi')\right| d\xi' \\
&\le \frac{1}{\sqrt{2\pi}^{n-1}} \int_{\operatorname{supp}(\hat{f})} e^{-|\xi'|x_n(\frac{1}{1+\delta})} e^{-|\xi'|x_n(\frac{\delta}{1+\delta})} |\hat{f}(\xi')|\, d\xi' \\
&\le e^{-\inf_{\xi'\in\operatorname{supp}(\hat{f})}(\frac{|\xi'|x_n}{1+\delta})} \frac{1}{\sqrt{2\pi}^{n-1}} \int_{\operatorname{supp}(\hat{f})} e^{-|\xi'|x_n(\frac{\delta}{1+\delta})} |\hat{f}(\xi')|\, d\xi' \\
&\le e^{-\rho|x_n|} \Big(\int |\hat{f}|^2\, d\xi'\Big)^{1/2} \Big(\int e^{-|\xi'|x_n(\frac{2\delta}{1+\delta})}\, d\xi'\Big)^{1/2} .
\end{aligned}$$

We have used the Cauchy–Schwarz inequality on the last line. We thus have the estimate on the decay of $u(x',x_n)$ for large x_n, namely

$$|u(x)| \le \frac{C(\delta)}{|x_n|^{(n-1)/2}} e^{-\rho|x_n|} \|f\|_{L^2} .$$

□

This result states that an electrostatic potential whose data on the boundary do not have support of its Fourier transform near $\xi' = 0$ does not penetrate the domain significantly, and therefore imaging techniques based on such potentials do not give rise to accurate renditions of details from deeper in a patient's body.

A similar bound holds for derivatives of $u(x)$ using the same lines of argument as in the proof above. This is again related to the interior regularity of solutions of elliptic equations. We will give a bound on multiple derivatives of $u(x)$. Using multi-index notation

$$\partial_x^\alpha u(x) = \partial_{x_1}^{\alpha_1} \dots \partial_{x_n}^{\alpha_n} u(x) ,$$

and assuming the hypotheses on the support of the Dirichlet data $f(x)$ as in Theorem 4.14, one follows a similar argument as in the proof above to show that

$$\begin{aligned}
|\partial_x^\alpha u(x',x_n)| &\le e^{-\rho} \int \Big(|\xi_1^{\alpha_1} \dots \xi_{n-1}^{\alpha_{n-1}}||\xi'|^{\alpha_n} |\hat{f}(\xi')| e^{-|\xi'|x_n(\frac{\delta}{1+\delta})}\Big)\, d\xi' \\
&\le \frac{C(\delta,\alpha)}{|x_n|^{\frac{n-1}{2}+|\alpha|}} e^{-\rho|x_n|} \|f\|_{L^2} .
\end{aligned}$$

4.9. Hadamard variational formula

It is rare to have such explicit formulae as (4.35) for the Dirichlet problem or (4.38) for the Dirichlet–Neumann operator, as is the case for the domain $\mathbb{R}^n_+$. More often one considers general domains, with less explicit solution procedures, so that it is reasonable to think to develop more general methods in order to understand the solution operators. In particular, it is relevant to ask whether the Green's function or the Poisson kernel vary continuously under perturbation of the domain. For the Dirichlet problem there is an elegant idea, originally due to Hadamard, to describe the Taylor expansion of the Green's function with respect to variations of the domain itself. This is described in the book of Garabedian [**8**] and is the topic of Project 4.15. Here we will describe this idea in a specific case, namely in terms of the Poisson kernel $\partial_N G(x', y)$ for domains which are perturbations of the upper half space $\mathbb{R}^n_+$.

Consider $\eta(x') \in C^1(\mathbb{R}^{n-1})$, whose graph defines the boundary of a domain $\Omega(\eta) = \{(x', x_n) \in \mathbb{R}^n : x_n > \eta(x')\}$. In this notation, $\Omega(0) = \mathbb{R}^n_+$. The domain $\Omega(\eta)$ has its Green's function $G_{\Omega(\eta)}(x', y)$, from which we obtain the Poisson kernel $\partial_N G_{\Omega(\eta)}(x', y) := D_{\Omega(\eta)}(x', y)$. The solution to the Dirichlet problem with Dirichlet data is thus given by the integral operator

$$u(x) = \mathbf{D}_{\Omega(\eta)} f(x) = \int_{\mathbb{R}^{n-1}} D_{\Omega(\eta)}(x', y) f(y)\, dS_y \ ,$$

where $dS_y = \sqrt{1 + |\nabla\eta|^2} dy'$. From (4.37) we see that $\mathbf{D}_{\Omega(0)} = e^{-x_n|D'|}$, a Fourier multiplier operator. In general $u(x) = \mathbf{D}_{\Omega(\eta)} f(x)$ is the bounded harmonic extension to the domain $\Omega(\eta)$ of the boundary data $f(x)$ defined on $\partial\Omega(\eta) = \{(x', x_n) : x_n = \eta(x')\}$. The solution operator $\mathbf{D}_{\Omega(\eta)}$ clearly depends upon the domain given by $\eta(x')$ in a nonlinear, global, and possibly complicated way. The Hadamard variational formula expresses the derivative of the operator $\mathbf{D}_{\Omega(\eta)}$ with respect to variations of η, giving a linear approximation to changes of the Green's function under perturbation of the domain.

Definition 4.15. Let $\mathbf{D}(\eta)$ be a bounded linear operator defined on L^2 that depends upon functions $\eta(x) \in C^1$. The bounded linear operator $\mathbf{A}(\eta)$ is the *Fréchet derivative* of $\mathbf{D}(\eta)$ with respect to $\eta(x)$ at the point $\eta = 0$ if it satisfies

$$\mathbf{A}(\lambda\eta) = \lambda\mathbf{A}(\eta) \quad \forall\lambda \in \mathbb{R} \ ,$$
$$\|\big(\mathbf{D}(\eta)f - \mathbf{D}(0)f\big) - \mathbf{A}(\eta)f\|_{L^2} \le C|\eta|^2_{C^1}\|f\|_{L^2} \ .$$

The notation that is used for the *variation*, or linear approximation, to $\mathbf{D}(\eta)$ is that $\mathbf{A}(\eta) = \delta\mathbf{D}(\eta)\big|_{\eta=0}$. This definition can of course be adapted to the more general situation of the operators $\mathbf{D}(\eta)$ mapping f in a Banach space

X to a Banach space Y, for η varying over a neighborhood of a third Banach space Z.

Without actually proving that $\mathbf{D}(\eta)$ is analytic with respect to $\eta \in C^1$ (which it is), we will derive a formula for the Fréchet derivative of $\mathbf{D}(\eta)$ for the domain that is the upper half space, for small $|\eta|_{C^1}$. A standard harmonic function on $\mathbb{R}^n_+$, and indeed on any of the domains $\Omega(\eta)$, is of course $\varphi_k(x) = e^{ik\cdot x'}e^{-|k|x_n}$ for each parameter $k \in \mathbb{R}^{n-1}$ fixed. Therefore

$$\mathbf{D}(0)\big(e^{ik\cdot x'}\big) = e^{ik\cdot x'}e^{-|k|x_n} = e^{-x_n|D'|}e^{ik\cdot x'} .$$

Given one of the domains $\Omega(\eta)$, the boundary values of $\varphi_k(x)$ on $\partial\Omega(\eta)$ are $e^{ik\cdot x'}e^{-|k|\eta(x')}$; therefore for $x_n > \eta(x')$, we know that

$$\mathbf{D}(\eta)\big(e^{ik\cdot x'}e^{-|k|\eta(x')}\big) = e^{ik\cdot x'}e^{-|k|x_n} .$$

Thus taking any point $(x', x_n) \in \Omega(0) \cap \Omega(\eta)$,

$$\begin{aligned}0 &= \mathbf{D}(\eta)\big(e^{ik\cdot x'}e^{-|k|\eta(x')}\big) - \mathbf{D}(0)\big(e^{ik\cdot x'}\big) \\ &= \mathbf{D}(\eta)\big((1 - \eta(x')|k| + \frac{1}{2}\eta^2(x')|k|^2 + \dots)e^{ik\cdot x'}\big) - \mathbf{D}(0)\big(e^{ik\cdot x'}\big) .\end{aligned}$$

This is to say that

$$\begin{aligned}&\mathbf{D}(\eta)\big(e^{ik\cdot x'}\big) - \mathbf{D}(0)\big(e^{ik\cdot x'}\big) - \mathbf{D}(\eta)\big(\eta(x')|k|e^{ik\cdot x'}\big) \\ &\quad = \mathbf{D}(\eta)\big((\frac{1}{2}\eta^2(x')|k|^2 + \dots)e^{ik\cdot x'}\big) .\end{aligned} \tag{4.40}$$

The facts are that the operator $\mathbf{D}(\eta)$ does have a Fréchet derivative $\mathbf{A}(\eta)$ at $\eta = 0$ and that the RHS is bounded by $C|\eta|^2_{C^1}|k|^2$ for small $|\eta|_{C^1}$. This allows us to compute $\mathbf{A}(\eta)$ from (4.40). Namely,

$$\mathbf{D}(\eta)(e^{ik\cdot x'}) - \mathbf{D}(0)(e^{ik\cdot x'}) = \mathbf{D}(0)\big(\eta(x')|D'|e^{ik\cdot x'}\big) + \mathcal{O}(|\eta|^2_{C^1}) ,$$

from which we read that

$$\mathbf{A}(\eta)e^{ik\cdot x'} = \mathbf{D}(0)\big(\eta(x')|D'|e^{ik\cdot x'}\big) = e^{-x_n|D'|}\big(\eta(x')|D'|\big)e^{ik\cdot x'} .$$

In other words, $\partial_\eta \mathbf{D}(\eta)\big|_{\eta=0} f = \mathbf{A}(\eta)f = e^{-x_n|D'|}\big(\eta(x')|D'|f\big)(x)$.

Now consider the Dirichlet–Neumann operator $\mathbf{G}(\eta)$ on domains $\Omega(\eta)$ which are perturbations of $\Omega(0) = \mathbb{R}^n_+$ and its Fréchet derivative at $\eta = 0$. Recall that

$$\mathbf{G}(\eta)f(x') = N_x \cdot \nabla u(x)\big|_{x_n=\eta(x')} ,$$

where $u(x)$ is the bounded harmonic extension of the Dirichlet data $f(x')$ to the domain $\Omega(\eta)$. Using again the family of harmonic functions $\varphi_k(x) = e^{ik\cdot x'}e^{-|k|x_n}$, we compute its boundary values $f(x')$ and its normal derivative

$N_x \cdot \nabla \varphi_k(x)$ on $\{x_n = \eta(x')\}$:

$$f(x') = e^{ik\cdot x'} e^{-|k|\eta(x')} = \big(1 - \eta(x')|k| + \mathcal{O}(|\eta|_{C^1}^2)\big) e^{ik\cdot x'} \ ,$$

$$\begin{aligned} \mathbf{G}(\eta) f(x') &= N_x \cdot \nabla \varphi_k(x) \\ &= \frac{1}{\sqrt{1+|\nabla\eta|^2}} (\partial_{x'}\eta, -1) \cdot (ik, -|k|) \nabla \varphi_k(x)\big|_{x_n=\eta(x')} \ . \end{aligned} \tag{4.41}$$

The Dirichlet–Neumann operator is not bounded on L^2, but it is bounded from $H^1(\mathbb{R}^{n-1})$ to L^2, as can be seen by its expression (4.38) as a Fourier multiplier when $\Omega = \mathbb{R}^n_+$. We seek the linear approximation to it among domain perturbations $\Omega(\eta)$ at the point $\eta = 0$. To calculate $\mathbf{B}(\eta) = \delta\mathbf{G}(\eta)\big|_{\eta=0} := \partial_\eta \mathbf{G}(\eta)\big|_{\eta=0}$, compare the first two terms of the LHS with the RHS of (4.41) in powers of η:

$$\begin{aligned} \mathbf{G}(\eta) f(x') &= \big(\mathbf{G}(0) + \mathbf{B}(\eta)\big) e^{ik\cdot x'} + \mathbf{G}(0)\big(-\eta(x')|k| e^{ik\cdot x'}\big) \\ &= |D'| e^{ik\cdot x'} + \mathbf{B}(\eta) e^{ik\cdot x'} - |D'|\big(\eta(x')|D'| e^{ik\cdot x'}\big) \end{aligned}$$

which is the LHS, and where the RHS is

$$\begin{aligned} N_x \cdot \nabla \varphi_k(x) &= |k| e^{ik\cdot x'} + \partial_{x'}\eta(x') \cdot ik\, e^{ik\cdot x'} - |k|^2 e^{ik\cdot x'} + \mathcal{O}(|\eta|_{C^1}^2) \\ &= |D'| e^{ik\cdot x'} + \partial_{x'}\big(\eta(x')\partial_{x'} e^{ik\cdot x'}\big) + \mathcal{O}(|\eta|_{C^1}^2) \ . \end{aligned}$$

Equating these expressions, solving for $\mathbf{B}(\eta)$, and applying the operators to a general $f(x')$ rather than the particular family of functions $e^{ik\cdot x'}$, we obtain

$$\mathbf{B}(\eta) f(x') = \partial_{x'}\big(\eta(x')\partial_{x'} f(x')\big) + |D'|\big(\eta(x')|D'| f(x')\big) \ . \tag{4.42}$$

Notice the symmetry under adjoints that is evident in this expression for $\mathbf{B}(\eta)$, reflecting the self-adjoint property of the Dirichlet–Neumann operator itself.

Exercises: Chapter 4

Exercise 4.1. Derive expression (4.36) for the case $n = 2$ from the Fourier integral, and show that $\omega_2 = 2\pi$. *Hint:* Complex variables techniques would be useful.

Exercise 4.2. In the case $n = 2$, the fundamental solution has the property that $\Delta\Gamma(|x-y|) = \delta_y(x)$. Show this by proving that the limit of the expression in (4.10) vanishes and that the limit in (4.11) holds.

Exercise 4.3 (Separation of variables)**.** Consider the boundary value problem for the domain $B_1(0) \subseteq \mathbb{R}^2$ for Dirichlet boundary data $u(x) = f(x)$ when $|x| = 1$:

$$\Delta u(x) = 0 \ , \qquad u(x) = f(x) \quad \text{for} \quad x \in \mathbb{S}_1(0) \ .$$

In polar coordinates $(r, \vartheta) \in \mathbb{R}^2$, Laplace's equation is written

$$\Delta u = (\partial_r^2 + \frac{1}{r}\partial_r)u + \frac{1}{r^2}\partial_\vartheta^2 u = 0 \ .$$

Special solutions can be obtained in the form $u(x) = v(r)w(\vartheta)$ so that

$$\partial_\vartheta^2 w(\vartheta) = -\lambda w(\vartheta) \ , \qquad (\partial_r^2 + \frac{1}{r}\partial_r)v(r) = \frac{\lambda}{r^2}v(r) \ .$$

(i) Show that this equation for $w(\vartheta)$ implies that

$$\lambda = k^2 \ , \qquad w(\vartheta) = c_k e^{ik\vartheta} + c_{-k}e^{-ik\vartheta} \quad \text{for} \quad k \in \mathbb{Z} \ .$$

(ii) Show that the resulting solutions $v(r)$ are therefore of the form

$$v(r) = r^k \quad \text{for} \quad k \geq 0 \ .$$

This uses the information that $u(x)$ is to be bounded at $r = 0$. Conclude that

$$u(r, \vartheta) = \sum_{k=0}^{+\infty} r^k \big(c_k e^{ik\vartheta} + c_{-k}e^{-ik\vartheta}\big)$$

is a harmonic function in $B_1(0)$.

(iii) Use this formula to prove that the Dirichlet problem for Laplace's equation on $B_1(0)$ has a solution for general Dirichlet data $f \in L^2(S_1(0))$.

Exercise 4.4. Give the solution to Laplace's equation on $B_1(0) \subseteq \mathbb{R}^2$ using the method of separation of variables for the following boundary data.

(1) $f(\vartheta) = a_1 \sin(\vartheta) + a_3 \sin(3\vartheta)$. What are the values of $u(0)$, $\partial_{x_1}u(0)$, and $\partial_{x_2}u(0)$?

(2) $f(\vartheta) = b_2 \cos(2\vartheta)$. What are the values of $u(0)$, $\partial_x u(0)$, and $\partial_x^2 u(0)$?

(3) $f(\vartheta) = 1 + 5\cos^3(\vartheta)$. What is the value of $u(0)$?

Exercise 4.5. Use the method of separation of variables on the square domain $\mathbb{D} := \{0 < x_1 < 2\pi \ , 0 < x_2 < 2\pi\}$ to solve the following boundary value problems for Laplace's equation

$$\Delta u = 0 \ .$$

(1) Dirichlet boundary data on the side $\{(x_1, 0) : 0 < x_1 < 2\pi\}$

$$f(x) = u(x_1, 0) = x_1(2\pi - x_1) \ ,$$

while on the remaining three sides set $f(x) = 0$.

(2) Dirichlet boundary data on opposite sides $\{(x_1, 0) : 0 < x_1 < 2\pi\}$ and $\{(x_1, 2\pi) : 0 < x_1 < 2\pi\}$

$$f(x) = u(x_1, 0) = x_1(2\pi - x_1) ,$$
$$f(x) = u(x_1, 2\pi) = -2x_1(2\pi - x_1) ,$$

while on the remaining two sides set $f(x) = 0$.

(3) Dirichlet boundary data on adjacent sides $\{(x_1, 0) : 0 < x_1 < 2\pi\}$ and $\{(0, x_2) : 0 < x_2 < 2\pi\}$

$$f(x) = u(x_1, 0) = 2\pi - x_1 , \quad f(x) = u(0, x_2) = 2\pi - x_2 ,$$

and $f(x) = 0$ on the remaining two sides.

Exercise 4.6 (Dirichlet Laplace operator on the square domain $\mathbb{D}$)**.** Eigenfunctions of the Laplace operator on a domain Ω with Dirichlet boundary conditions satisfy the equation

$$\Delta u = \lambda u , \quad u(x) = 0 \quad \text{for} \quad x \in \partial\mathbb{D} ,$$

and λ is the associated eigenvalue. In the case that $\Omega = \mathbb{D}$ is the square, eigenfunction-eigenvalue pairs can be expressed by separation of variables methods.

(1) Use the Ansatz $u(x) = v(x_1)w(x_2)$ to solve for eigenfunctions and eigenvalues of the Dirichlet Laplacian on the square $\mathbb{D}$ and their associated eigenvalues.

(2) Show that all of the eigenvalues satisfy $\lambda < 0$ and that the eigenfunction $u(x)$ associated with the eigenvalue λ that is smallest in absolute value satisfies

$$u(x) \neq 0 \quad \text{for} \quad x \in \mathbb{D} .$$

(3) Show that this set of eigenfunctions is complete, meaning that linear combinations of them are dense in $L^2(\mathbb{D})$. Enumerate the eigenfunction-eigenvalue pairs $(u_k(x), \lambda_k)$ as you prefer, and normalize the eigenfunctions so that

$$\int_{\mathbb{D}} |u_k(x)|^2 \, dx = 1 , \quad \int_{\mathbb{D}} u_k(x) u_\ell(x) \, dx = 0$$

for $k \neq \ell$.

(4) Use an eigenfunction expansion to give an expression for the Green's function for $\mathbb{D}$ in the form

$$G(x, y) = \sum_k \frac{1}{\lambda_k} u_k(x) u_k(y) .$$

Now show that a solution to the Poisson problem

$$\Delta u(x) = h(x) \ , \quad x \in \mathbb{D} \ , \quad u(x) = 0 \ , \quad x \in \partial\mathbb{D}$$

can be expressed by

$$u(x) = \int_{\mathbb{D}} G(x,y)h(y)\, dy \ .$$

Exercise 4.7 (Mixed boundary conditions)**.** Consider the domain $\Omega = B_1(0) \cap \mathbb{R}^2_+$; this is a half disk. On the boundary $\partial\Omega \cap \{x_2 = 0\}$, give the following boundary conditions for Laplace's equation:

$$u(x_1, 0) = 0 \quad \text{for} \quad x_1 > 0 \ , \quad \partial_{x_2} u(x_1, 0) = 0 \quad \text{for} \quad x_1 < 0 \ .$$

On the boundary $\{|x| = 1,\, x_2 > 0\}$, specify that $u(x) = \sin(\vartheta/2)$. Show that $u(x) = \sqrt{r}\sin(\vartheta/2)$ satisfies Laplace's equation and these boundary conditions. However, show that $u(x)$ has a square root singularity on the axis $\{x_2 = 0\}$ even though the boundary data is smooth.

Exercise 4.8. Prove the second version of Gauss' law of arithmetic mean (4.14) for harmonic functions,

$$u(y) = \frac{n}{\omega_n \rho^n} \int_{B_\rho(y)} u(x)\, dx \ .$$

Exercise 4.9. Prove that a function $u(x) \in C^2(\Omega)$ that satisfies

$$\Delta u \geq 0 \tag{4.43}$$

is subharmonic in Ω in the sense of Definition 4.7. Prove that if $u(x) \in C^2(\Omega)$ is subharmonic, then it satisfies inequality (4.43).

Exercise 4.10. Show that the Green's function $G(x,y)$ on a domain $\Omega \subseteq \mathbb{R}^n$ satisfies the property of symmetry $G(x,y) = G(y,x)$. Conclude that the resulting operator Δ^{-1} with Dirichlet boundary conditions is self-adjoint on $L^2(\Omega)$.

Exercise 4.11. Prove the geometric property expressed by (4.22) for points x, y such that $x \in \mathbb{S}^{n-1}_R$ and $y \in B_R(0)$, with $y^* = \frac{R^2}{|y|^2} y$, the inversion of y in the sphere $\mathbb{S}^{n-1}_R$. That is, for any $y \notin \mathbb{S}^{n-1}_R$, the sphere $\mathbb{S}^{n-1}_R$ is the locus of points x such that the ratio

$$\frac{|x - y^*|}{|x - y|} = \frac{R}{|y|}$$

is constant.

Exercise 4.12. Calculate the normal derivative of the Green's function (4.23) for the domain $B_R(0)$ with respect to x on the boundary $x \in \mathbb{S}^{n-1}_R$, verifying expression (4.26).

Exercise 4.13. Show that in $\mathbb{R}^2$, the Laplace operator is invariant under conformal transformations. That is, if $x = x_1 + ix_2$ and $y = y_1 + iy_2$ satisfy $x = \tau(y)$ for an analytic function τ, then under the change of variables $u(x) \to w(y) = u(\tau(y))$ the Laplacian is invariant, namely

$$\Delta_x u = \Delta_y w \ .$$

Hint: One approach to this question is to use the fact that the Laplace operator on a domain $\Omega_1 \subseteq \mathbb{R}^2$ is the gradient of the energy functional

$$L(u) = \frac{1}{2} \int_{\Omega_1} |\partial_x u(x)|^2 \, dx \ .$$

Namely, $\mathrm{grad}L(u) = -\Delta u$ in the sense that

$$\frac{d}{d\sigma}\Big|_{\sigma=0} L(u + \sigma v) = \int_{\Omega_1} \mathrm{grad}\, L(u)\, v \, dx = \int_{\Omega_1} -\Delta u \, v \, dx$$

for all compactly supported test functions $v \in C^2(\Omega_1)$. Under a change of variables $x = \tau(y)$ with $w(y) = u(\tau(y))$, one has

$$|\partial_y w(y)|^2 = \langle \partial_x u, \ (\partial_y \tau)^T (\partial_y \tau) \partial_x u \rangle \ , \qquad dx_1 dx_2 = \det(\partial_y \tau) dy_1 dy_2 \ .$$

It suffices to show that for analytic transformations $x = \tau(y)$,

$$(\partial_y \tau)^T (\partial_y \tau) = \det(\partial_y \tau) I \ .$$

These expressions appear when transforming the energy functional $L(u)$ over $x \in \Omega_1$ to $L(w)$ over $y \in \Omega_2 = \tau^{-1}(\Omega_1)$.

Exercise 4.14. For simply connected domains Ω in $\mathbb{R}^2$, describe the Green's function in terms of the Riemann map of Ω to $\mathbb{R}^2_+$.

Projects: Chapter 4

Project 4.15 (Hadamard variational formula)**.** This project is an extension of the discussion in section 4.9 on perturbations of harmonic functions with respect to their domain of definition. In particular, it studies perturbations of the Green's function of the domain. The goal is to prove the formula that under perturbation of a domain Ω by small displacements $\eta(x)$ in the normal direction, the Green's function for the domain, $G(x,y)$ for $x, y \in \Omega$, satisfies the variational formula

$$\delta G(x,y) = -\int_{\partial\Omega} N_z \cdot \nabla_z G(x,z)\, \eta(z)\, N_z \cdot \nabla_z G(z,y)\, dS_z \ . \tag{4.44}$$

This formula was derived by J. Hadamard in 1915; it is from one point of view a version of the resolvant identity.

Consider a compact domain $\Omega \subseteq \mathbb{R}^n$ with real analytic boundary, and a nearby domain $\Omega_1 \subseteq \mathbb{R}^n$ which is a real analytic perturbation of Ω, in the sense that the boundary $\partial\Omega_1$ can be parametrized by $x_1 \in \partial\Omega_1$, where

$x_1 = x + \eta(x)N$ for $x \in \partial\Omega$. The domain Ω and the perturbation η are assumed to be real analytic, and considering η over a small complex neighborhood of $\partial\Omega$, the norm $\|\eta\|_{L^\infty}$ is assumed to be sufficiently small so that any harmonic function on Ω can be extended to a harmonic function in a full neighborhood of Ω, and similarly for Ω_1. Actually, formula (4.44) applies in more generality than this; a derivation in the case of domains Ω with boundaries that are C^2 was derived by Garabedian and Schiffer, and is described in Garabedian [**8**].

(i) Let Ω and Ω_1 be two domains, and let $x, y \in \Omega \cap \Omega_1$ be two points interior to both domains. The Green's functions $G(x, y)$ for Ω and $G_1(x, y)$ for Ω_1 are both well defined. Furthermore, suppose that $\eta(x)$ is sufficiently small so that the harmonic extension of $G(x, y)$ in y to a neighborhood of Ω, for x fixed, is defined on Ω_1, and vice versa for $G_1(x, y)$.

Use Green's identity as in the proof of Theorem 4.3 to show that

$$\begin{aligned}\int_\Omega & \nabla_z G(x,z) \cdot \nabla_z G_1(z,y)\, dz \\ &= \int_{\partial\Omega} G(x,z)\, N_z \cdot G_1(z,y)\, dS_z - G(x,y) = -G(x,y) \ , \\ \int_{\Omega_1} & \nabla_z G(x,z) \cdot \nabla_z G_1(z,y)\, dz \\ &= \int_{\partial\Omega_1} N_z \cdot G(x,z)\, G_1(z,y)\, dS_z - G_1(x,y) = -G_1(x,y) \ .\end{aligned}$$

Conclude that

$$G_1(x,y) - G(x,y) = -\int_{\Omega_1} \nabla G(x,z) \cdot \nabla G_1(z,y)\, dz + \int_\Omega \nabla G(x,z) \cdot \nabla G_1(z,y)\, dz \ .$$

(ii) Consider the above expression as an integral of $\nabla_z G(x,z) \cdot \nabla_z G_1(z,y)$ over the difference between the two domains (using the sign convention that one adds if $z \in \Omega_1 \backslash \Omega$ and subtracts if $z \in \Omega \backslash \Omega_1$). Because $\nabla_z G(x,z) \cdot T = 0$ for all tangent vectors to the boundary $\partial\Omega$, show that

$$\begin{aligned}\int_{\Omega_1 - \Omega} & \nabla G(x,z) \cdot \nabla G(z,y)\, dz \\ &= \int_{\partial\Omega} N_z \cdot \nabla_z G(x,z) \eta(z) N_z \cdot \nabla_z G_1(z,y)\, dz + \mathcal{O}(|\eta|^2_{L^\infty}) \ .\end{aligned}$$

Therefore

$$\begin{aligned} G_1(x,y) - G(x,y) &= -\int_{\Omega_1 - \Omega} \nabla G(x,z) \cdot \nabla G_1(z,y)\, dz \\ &= -\int_{\partial\Omega} N_z \cdot \nabla_z G(x,z) \eta(z) N_z \cdot \nabla_z G_1(z,y)\, dz + \mathcal{O}(|\eta|^2_{L^\infty}) \ . \qquad (4.45)\end{aligned}$$

(iii) Finally, use formula (4.45) again in order to show that

$$G_1(z,y) - G(z,y) \sim \mathcal{O}(|\eta|_{L^\infty}) \ ,$$

therefore

$$G_1(x,y) - G(x,y) = -\int_{\partial\Omega} N_z \cdot \nabla_z G(x,z)\eta(z) N_z \cdot \nabla_z G(z,y)\, dz + \mathcal{O}(|\eta|^2_{L^\infty}) \ ,$$

which is a statement of the variational formula of Hadamard.

(iv) For a more open-ended project, also derive the variational formula of Hadamard for general C^2 smooth domains Ω and C^2 domain perturbations $\eta(x)$. A useful reference is Garabedian [**8**].

Chapter 5

Properties of the Fourier transform

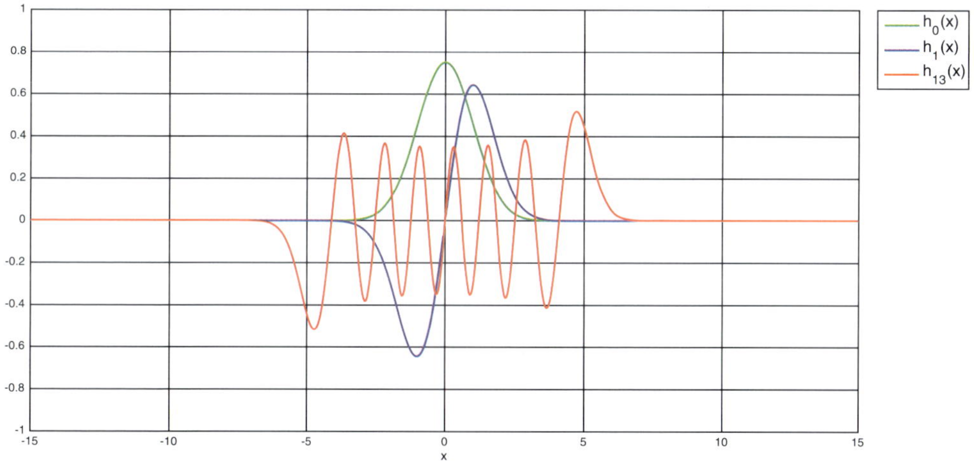

Normalised Hermite function $h_n(x) = \frac{1}{(2^n\sqrt{\pi}n!)^{\frac{1}{2}}} H_n(x)e^{-\frac{x^2}{2}}$, the graphs of h_1, h_2, and h_{13}.

The purposes of this section are to raise our level of sophistication of the analysis of the Fourier transform and to make up our "backlog" of analytic justification of our work in the previous several sections. The Fourier transform plays a very important role in analysis, and for this reason it has been thoroughly analyzed from many points of view and by many people in many different settings. We will work through the content of three principal settings in this chapter, the ones most important to analysis and PDE.

The cases are

(1) Hilbert spaces, covered in section 5.1,

(2) Schwartz class, covered in section 5.2, and

(3) L^1 integrability, covered in section 5.3.

5.1. Hilbert spaces

Definition 5.1. The usual Hilbert space for us is $L^2(\mathbb{R}^n)$, which consists of the square-integrable measurable complex valued functions on $\mathbb{R}^n$:

$$L^2(\mathbb{R}^n) = \{f \text{ Lebesgue measurable in } \mathbb{R}^n, \int_{\mathbb{R}^n} |f(x)|^2 dx < +\infty\}.$$

Proposition 5.2 (Elementary properties of $L^2(\mathbb{R}^n)$)**.** (1) *The space of functions $L^2(\mathbb{R}^n)$ is a linear space; namely, if $f, g \in L^2$, then this implies that $af + bg \in L^2$ for all $a, b \in \mathbb{C}$.*

(2) *The topology on $L^2(\mathbb{R}^n)$ is given by a norm*

$$\|f\|_{L^2} = \left(\int_{\mathbb{R}^n} |f(x)|^2\, dx\right)^{1/2}, \quad \|f\|_{L^2} = 0 \Leftrightarrow f = 0\ .$$

Norms are homogeneous of degree 1*:*

$$\|\alpha f\|_{L^2} = |\alpha| \|f\|_{L^2}\ .$$

(3) *Norms satisfy the triangle (or Minkowski) inequality:*

$$\|f + g\|_{L^2} \le \|f\|_{L^2} + \|g\|_{L^2}\ .$$

(4) *The linear space $L^2(\mathbb{R}^n)$ is* complete *with respect to the norm $\|f\|_{L^2}$. That is, if $\{f_n\}_{n=1}^\infty \subseteq L^2(\mathbb{R}^n)$ is a Cauchy sequence of functions, then there exists a limit function $f(x) \in L^2(\mathbb{R}^n)$ such that $f_n(x) \to_{L^2} f(x)$ (meaning $\|f_n(x) - f(x)\|_{L^2} \to 0$).*

(5) *Furthermore, the norm on $L^2(\mathbb{R}^n)$ is given by an inner product:*

$$\|f\|_{L^2}^2 = \langle f, f\rangle \quad \textit{where} \quad \langle f, g\rangle = \int f(x)\overline{g(x)}\, dx\ . \tag{5.1}$$

The inner product satisfies the Cauchy–Schwarz inequality:

$$|\langle f, g\rangle| \le \|f\|_{L^2}\|g\|_{L^2}\ .$$

(6) *The "dual space", or space of bounded linear functionals on $L^2(\mathbb{R}^n)$, is $L^2(\mathbb{R}^n)$ itself.*

Definition 5.3 (Banach and Hilbert spaces)**.** A linear space with a norm, which is complete, is a *Banach space.* A Banach space whose norm is given by an inner product is a *Hilbert space.*

Our favorite Hilbert space $L^2(\mathbb{R}^n)$ has its inner product defined in (5.1). One reason that $L^2(\mathbb{R}^n)$ is a natural setting for the Fourier transform is that it is preserved under the transform. In $\mathbb{R}^n$, we define the transform as follows. For $\xi \in \mathbb{R}^n$,

$$\hat{f}(\xi) = \frac{1}{\sqrt{2\pi}^n} \int_{\mathbb{R}^n} e^{-i\xi\cdot x} f(x)dx = (\mathcal{F}f)(\xi) .$$

Theorem 5.4. *The Fourier transform is an isometry of $L^2(\mathbb{R}^n)$. That is, $\|f\|_{L^2(\mathbb{R}^n_x)} = \|\hat{f}\|_{L^2(\mathbb{R}^n_\xi)}$, or, stated in other words,*

$$\int_{\mathbb{R}^n_x} |f(x)|^2 \, dx = \int_{\mathbb{R}^n_\xi} |\hat{f}(\xi)|^2 \, d\xi .$$

This equality between the L^2-norms of a function and its Fourier transform is known as the *Plancherel identity*; it is a general fact about the Fourier transform that holds in many settings. The proof of Theorem 5.4 is deferred until the end of our discussion of Schwartz class.

Other examples of Hilbert spaces and Banach spaces as tools of analysis include the following.

(1) Sobolev spaces $H^s(\mathbb{R}^n)$ (Hilbert spaces based on L^2-norms): For $s \in \mathbb{N}$,

$$H^s(\mathbb{R}^n) = \{f(x) \, : \, \sum_{0\le|\alpha|\le s} \Big(\int |\partial_x^\alpha f(x)|^2 \, dx\Big)^{\frac{1}{2}} < +\infty\} .$$

The norm of a function f in this Sobolev space measures the size of derivatives of f as well as the size of f itself. We are using our previously introduced notation that a convenient device for multivariable objects involves multi-indices α, β: $\alpha = (\alpha_1, \alpha_2, \ldots, \alpha_n)$, $\beta = (\beta_1, \beta_2, \ldots, \beta_n)$, with each $\alpha_i, \beta_j \in \mathbb{N}$. The order of a multi-index is given by $|\alpha| = |\alpha_1| + \cdots + |\alpha_n|$ so that $x^\alpha = x_1^{\alpha_1} x_2^{\alpha_2} \ldots x_n^{\alpha_n}$ is a monomial of degree $|\alpha|$ and $\partial_x^\beta = \partial_{x_1}^{\beta_1} \partial_{x_2}^{\beta_2} \ldots \partial_{x_n}^{\beta_n} = \Pi \partial_{x_j}^{\beta_j}$ is a differential operator of order $|\beta|$. The family $H^s(\mathbb{R}^n)$ is a *scale of spaces* which are nested in terms of decreasing index s: $H^s \subseteq H^{s-1} \subseteq \cdots \subseteq L^2(\mathbb{R}^n)$.

(2) L^p spaces (Banach spaces): For $1 \le p < +\infty$,

$$L^p(\mathbb{R}^n) = \{f(x) \, : \, \Big(\int_{\mathbb{R}^n} |f(x)|^p \, dx\Big)^{\frac{1}{p}} < +\infty\} .$$

For $p = \infty$, the space of bounded measurable functions on $\mathbb{R}^n$ is denoted by $L^\infty(\mathbb{R}^n)$ and has a norm given by

$$\|f\|_{L^\infty} := \sup_{x\in\mathbb{R}^n} |f(x)| .$$

(3) Sobolev spaces $W^{s,p}(\mathbb{R}^n)$ (Banach spaces modeled on L^p-norms, which again measure derivatives of f):

$$W^{s,p}(\mathbb{R}^n) = \{f \,:\, \sum_{0\le|\alpha|\le s} \Big(\int |\partial_x^\alpha f(x)|^p \, dx \Big)^{\frac{1}{p}} < +\infty\} \ .$$

(4) Schauder spaces $C^{s,\gamma}(\Omega)$ (Banach spaces modeled on C^0-norms): For $\Omega \subseteq \mathbb{R}^n$ a domain, that is, an open set with smooth boundary,

$$C^{s,\gamma}(\Omega) = \{f(x) \,:\, \sup_{x\in\Omega} |\partial_x^\alpha f(x)| < +\infty \ \forall |\alpha| \le s$$
$$\text{and } \forall |\alpha| = s, \quad \sup_{x,y\in\Omega, x\ne y} \frac{|\partial_x^\alpha f(x) - \partial_x^\alpha f(y)|}{|x-y|^\gamma} < +\infty\} \ .$$

The space of Lipschitz functions is in this class; namely, $f \in Lip(\mathbb{R}^n) = C^{0,1}(\mathbb{R}^n)$ when there is a constant C such that

$$|f(x) - f(y)| \le C|x-y| \quad \text{for all} \quad x,y \in \mathbb{R}^n \ .$$

The many relationships of inclusion among these spaces, and their associated inequalities of norms, are important information for the analysis of PDEs. A principal one that is very often used in PDEs is the Sobolev Embedding Theorem. A version of this result is as follows.

Theorem 5.5 (Sobolev embedding). *If $f(x) \in H^s(\mathbb{R}^n)$ for a Sobolev index $s > \frac{n}{2}$, then*

$$|f(x)|_{L^\infty} \le C_n \|f\|_{H^s} \ .$$

In other words $H^s \subseteq L^\infty$ for $s > \frac{n}{2}$, and there is a bound on the inclusion operator $\iota : H^s \to L^\infty$ given by the constant C_n.

Proof. Consider the value of $f(x)$ at an arbitrary point of $\mathbb{R}^n$,

$$|f(x)| = \Big|\frac{1}{\sqrt{2\pi}^n} \int_{\mathbb{R}^n} e^{i\xi\cdot x} \hat{f}(\xi) \, d\xi\Big|$$
$$= \Big|\frac{1}{\sqrt{2\pi}^n} \int_{\mathbb{R}^n} e^{i\xi\cdot x} \frac{1}{(1+|\xi|^2)^{\frac{s}{2}}} (1+|\xi|^2)^{\frac{s}{2}} \hat{f}(\xi) \, d\xi\Big| \ .$$

Using the Cauchy–Schwarz inequality on this last expression,

$$|f(x)| \le \frac{1}{\sqrt{2\pi}^n} \Big(\int \frac{1}{(1+|\xi|^2)^s} \, d\xi \Big)^{\frac{1}{2}} \Big(\int (1+|\xi|^2)^s |\hat{f}(\xi)|^2 \, d\xi \Big)^{\frac{1}{2}} \ . \tag{5.2}$$

Because derivatives are transformed to multiplications by $i\xi$ under the Fourier transform, by the binomial theorem,

$$\Big(\int (1+|\xi|^2)^s |\hat{f}(\xi)|^2\, d\xi\Big)^{\frac{1}{2}} = \Big(\sum_{j=0}^{s} \int \binom{s}{j} |\xi|^{2j} |\hat{f}(\xi)|^2\, d\xi\Big)^{\frac{1}{2}}$$

$$= \Big(\sum_{j=0}^{s} \int \binom{s}{j} |\nabla^j f(x)|^2\, dx\Big)^{\frac{1}{2}} \le C_n \|f\|_{H^s} ,$$

where we have used the Plancherel identity for $f(x)$ and its derivatives in the last line. This holds of course for any integer s. It remains to bound the RHS of (5.2),

$$\Big(\int \frac{1}{(1+|\xi|^2)^s}\, d\xi\Big)^{\frac{1}{2}} = \Big(\int_{\mathbb{S}^{n-1}} \int_0^{+\infty} \frac{1}{(1+r^2)^s} r^{n-1} dr\, dS_\varphi\Big)^{\frac{1}{2}} ,$$

and if $s > \frac{n}{2}$, then this quantity is finite, which finishes the proof. □

5.2. Schwartz class

This linear space of functions was introduced by Laurent Schwartz (École Polytechnique, Paris) in order to embody a very convenient class of functions with which to work, for which one can integrate by parts and interchange integrations with differentiations with impunity. Colloquially, $f(x)$ is a Schwartz class function if it is infinitely differentiable and if it decays along with all of its derivatives as $|x| \to \infty$ faster than any polynomial.

Recall our previously introduced notation that is a convenient device for multivariable objects: the quantities α, β are multi-indices,

$$\alpha = (\alpha_1, \alpha_2, \ldots, \alpha_n), \quad \text{and} \quad \beta = (\beta_1, \beta_2, \ldots, \beta_n) .$$

Definition 5.6 (Schwartz class). The Schwartz class consists of those functions $f(x)$ defined over $\mathbb{R}^n$ such that for all α, β,

$$\sup_{x\in\mathbb{R}^n} |x^\alpha \partial_x^\beta f(x)| < +\infty.$$

It is clearly a linear space of functions and is denoted by $\mathcal{S}$.

The multi-index notation is a device for describing many combinatorially complex objects, which can be written very conveniently. For example, Taylor series in n dimensions can be stated

$$f(x) \sim \sum_\alpha \frac{1}{\alpha!} \partial_x^\alpha f(y)(x-y)^\alpha$$

with $\alpha! = \alpha_1!\alpha_2!\ldots\alpha_n!$.

Here are several examples of particular functions where we can compare their properties of smoothness and decay in $x \in \mathbb{R}^n$ to those of Schwartz class:

$$
\begin{aligned}
&f(x) = e^{-\frac{x^2}{2}} \in \mathcal{S} \ , \\
&f(x) = (1+|x|^2)^{-\frac{b}{2}} \notin \mathcal{S} \text{ but } f(x) \in L^2(\mathbb{R}^n) \text{ for } b > n/2 \ , \\
&f(x) = e^{-|x|} \notin \mathcal{S} \text{ for a different reason } .
\end{aligned}
$$

Every Schwartz class function $f(x)$ is also in $L^2(\mathbb{R}^n)$, as it is bounded and decays rapidly. Furthermore, Schwartz class is a linear subspace of $L^2(\mathbb{R}^n)$ which is dense in the L^2 topology. That is, for any $f \in L^2(\mathbb{R}^n)$ there is a sequence $\{f_n\}_{n=1}^{\infty} \subseteq \mathcal{S}$ such that $\lim_{n\to\infty} \|f_n - f\|_{L^2} = 0$. The proof of this fact appears in the exercises.

The space C_0^∞ of infinitely differentiable functions with compact support is a subset of $\mathcal{S}$ as a subspace, $C_0^\infty \subseteq \mathcal{S}$, and the space C^∞ of infinitely differentiable functions contains Schwartz class, $\mathcal{S} \subseteq C^\infty$. The space $C^\omega(\mathbb{R}^n)$ of *real analytic functions* on $\mathbb{R}^n$ (the restriction to $x \in \mathbb{R}^n$ of functions that are analytic in a neighborhood of $\mathbb{R}^n \subseteq \mathbb{C}^n$) is a subset of C^∞, but not every C^∞ function is analytic.

The topology of $\mathcal{S}$ is defined using a countable family of *seminorms*

$$
\|f\|_{\alpha,\beta} = \sup_{x\in\mathbb{R}^n} |x^\alpha \partial_x^\beta f(x)| \ .
$$

That is to say, a sequence $\{f_n\}_{n=1}^{\infty}$ converges to f if and only if for all multi-indices α, β, the seminorms

$$
\|f_n - f\|_{\alpha,\beta} = \sup_x |x^\alpha \partial_x^\beta (f_n(x) - f(x))| \to 0
$$

with $n \to \infty$. No uniformity in α, β is implied, however. Some functional analytic remarks are in order at this point. The space $\mathcal{S}$ is a linear space but not a Banach space, as there is no way to describe this sense of convergence by using a norm. It is, however, a metric space with the distance between two points given by

$$
\rho(f,g) = \sum_{\alpha,\beta} \frac{1}{2^{n(|\alpha|+|\beta|)}} \frac{\|f-g\|_{\alpha,\beta}}{1+\|f-g\|_{\alpha,\beta}} \ ,
$$

and it is complete with respect to this metric. Clearly $\{f_n(x)\}_{n=1}^{\infty}$ converges in $\mathcal{S}$ to $f(x)$ if and only if $\rho(f_n, f) \to 0$ if and only if all $\|f_n - f\|_{\alpha,\beta} \to 0$ as $n \to \infty$. The notation for limits of sequences of functions in this Schwartz class sense is to write $\mathcal{S} - \lim f_n = f$. Metric spaces such as this one whose topology is given by a countable family of seminorms are called *Fréchet spaces*.

Lemma 5.7 (Technical lemma). *This result quantifies the properties of scaling limits in two cases, and these limits give rise to well-defined functionals on $\mathcal{S}$.*

(i) (*Approximations of the identity*). *Let $g \in \mathcal{S}$ with $g(0) = 1$. Then for all $f \in \mathcal{S}$,*

$$\mathcal{S} - \lim_{\varepsilon \to 0} (g(\varepsilon x) f(x)) = f(x) \ .$$

(ii) (*Approximations of the Dirac δ-function*). *Let $h(x) \in L^1(\mathbb{R}^n)$ with $\int h(x)dx = 1$, and suppose that $f(x)$ is bounded, and continuous at $x = 0$. Then*

$$\lim_{\varepsilon \to 0} \int f(x) \frac{1}{\varepsilon^n} h(\frac{x}{\varepsilon}) dx = f(0) \ .$$

For $h(x) \in L^1(\mathbb{R}^n)$ with $\int h(x)\, dx = 1$, the limit operation in (5.7) describes the approximation of the Dirac δ-function by rescaling a suitably normalized L^1 function $h(x)$.

Proof of Lemma 5.7(ii). By change of variables,

$$\int \frac{1}{\varepsilon^n} h(\frac{x}{\varepsilon})\, dx = \int h(x')\, dx' = 1$$

with $x' = \frac{x}{\varepsilon}$. Given $\delta > 0$, choose $r > 0$ such that $|f(x) - f(0)| < \delta$ for all $|x| < r$. Then write

$$\int f(x) \frac{1}{\varepsilon^n} h(\frac{x}{\varepsilon})\, dx = \int f(0) \frac{1}{\varepsilon^n} h(\frac{x}{\varepsilon})\, dx + \int (f(x) - f(0)) \frac{1}{\varepsilon^n} h(\frac{x}{\varepsilon})\, dx \ .$$

The first term is $f(0)$. Split the second term in two,

$$\begin{aligned} \Big|\int_{|x|<r} (f(x) - f(0)) \frac{1}{\varepsilon^n} h(\frac{x}{\varepsilon})\, dx\Big| &\le \delta \Big|\int_{|x|<r} \frac{1}{\varepsilon^n} h(\frac{x}{\varepsilon})\, dx\Big| \le \delta \|h\|_{L^1} \ , \\ \Big|\int_{|x|\ge r} (f(x) - f(0)) \frac{1}{\varepsilon^n} h(\frac{x}{\varepsilon})\, dx\Big| &\le 2 \sup_x |f(x)| \int_{|x|\ge r} \frac{1}{\varepsilon^n} |h(\frac{x}{\varepsilon})|\, dx \\ &= 2 \sup_x |f(x)| \int_{|x'|\ge \frac{r}{\varepsilon}} |h(x')| dx' \ , \end{aligned}$$

and by the Lebesgue dominated convergence theorem, the RHS of the last term with r fixed tends to zero with $\varepsilon \to 0$. □

The proof of Lemma 5.7(i) appears in the exercises of this chapter.

Corollary 5.8. *The space $C_0^\infty \subseteq \mathcal{S}$ is a dense subspace of $\mathcal{S}$.*

Proof. This uses Lemma 5.7: Take any $g \in C_0^\infty$ with $g(0) = 1$. Then for $f \in \mathcal{S}$ arbitrary,

$$g(\varepsilon x) f(x) \in C_0^\infty$$

and

$$\mathcal{S} - \lim_{\varepsilon \to 0} (g(\varepsilon x) f(x)) = f(x) \ .$$

□

The space of *distributions* is the space of continuous linear functionals on C_0^∞, $\mathcal{D}' = (C_0^\infty)'$, otherwise known as its dual space. The space of *tempered distributions* is the dual space of $\mathcal{S}$. Also, we write $\mathcal{E}'$ for the dual space to C^∞, the space of infinitely differentiable functions. Since $C_0^\infty \subseteq \mathcal{S} \subseteq C^\infty$, then $\mathcal{E}' \subseteq \mathcal{S}' \subseteq \mathcal{D}'$. An example of a distribution is the Dirac δ-function, $\delta(f) := f(0)$, which is a distribution in $\mathcal{E}'$ and hence in all of the above spaces of continuous linear functionals.

The Fourier transform acts nicely on $\mathcal{S}$; for $f(x) \in \mathcal{S}$ define the Fourier transform as usual:

$$\hat{f}(\xi) = \frac{1}{\sqrt{2\pi}^n} \int e^{-i\xi \cdot y} f(y)\, dy = \mathcal{F}(f)(\xi) \ .$$

One checks that this integral converges absolutely because $f(x) \in \mathcal{S}$; indeed

$$|e^{-i\xi \cdot x} f(x)| \le \|f\|_{N,0} \frac{1}{(1+|x|^2)^{\frac{N}{2}}} \ ,$$

and $N > n+1$ will do for this. For Schwartz class $f(x)$ we may exchange integrations and differentiations as much as we wish; therefore the following list of properties holds for $\hat{f}(\xi)$.

Proposition 5.9 (Elementary properties of the Fourier transform on $\mathcal{S}$).
(1) $\widehat{\partial_x f}(\xi) = (i\xi)\hat{f}(\xi)$.

(2) $\widehat{xf}(\xi) = i\partial_\xi \hat{f}(\xi)$.

Notice the vectorial notation; x and $i\partial_\xi$ are vector operations.

Define two operations on functions: $\tau_h f(x) = f(x-h)$ (translations) and $(\sigma_\lambda f)(x) = f(\lambda x)$ (dilations).

(3) $\widehat{\tau_h f}(\xi) = e^{-ih\xi}\hat{f}(\xi)$, *a* position boost.

(4) $\widehat{e^{ihx} f}(\xi) = \hat{f}(\xi - h)$, *a* momentum boost.

(5) $\widehat{\sigma_\lambda f}(\xi) = \frac{1}{|\lambda|^n}\sigma_{(1/\lambda)}(\hat{f})(\xi) = \frac{1}{|\lambda|^n}\hat{f}(\xi/\lambda)$.

The principal reason why Schwartz class is well suited for Fourier analysis is that it is invariant under the Fourier transform.

Theorem 5.10. *The Fourier transform maps $\mathcal{S}$ into itself; that is, whenever $f \in \mathcal{S}$, then $\hat{f} = \mathcal{F}(f) \in \mathcal{S}$.*

Proof. For $f(x) \in \mathcal{S}$, at the very least $\hat{f}(\xi)$ makes sense as a convergent integral. In order to check that $\hat{f}(\xi) \in \mathcal{S}$, we have to check that all of the seminorms are finite, which we do using the properties described in Proposition 5.9:

$$\begin{aligned}\|\hat{f}\|_{\alpha,\beta} &= \sup_{\xi} |\xi^{\alpha}\partial_{\xi}^{\beta}\hat{f}(\xi)| \\ &= \sup_{\xi}\Big|\frac{1}{\sqrt{2\pi}^{n}}\int e^{-i\xi\cdot x}((\frac{1}{i}\partial_x)^{\alpha}(\frac{1}{i}x)^{\beta}f(x))\,dx\Big|.\end{aligned}$$

This integrand is still bounded by $C_{\beta+N,\alpha}(1+|x|^2)^{-\frac{N}{2}}$, and hence the supremum over $\xi \in \mathbb{R}^n$ is finite. Furthermore, the Fourier transform is continuous on $\mathcal{S}$. Suppose that $\{f_n(x)\}_{n=1}^{\infty}$, and $f_n \underset{\mathcal{S}}{\to} f$. This implies that for each α, β,

$$\|\hat{f}_n - \hat{f}\|_{\alpha,\beta} = \sup_{\xi}\Big|\frac{1}{\sqrt{2\pi}^{n}}\int e^{-i\xi\cdot x}((\frac{1}{i}\partial_x)^{\alpha}(\frac{1}{i}x)^{\beta}(f_n(x)-f(x)))\,dx\Big| \to 0$$

with $n \to \infty$ as well. Hence $\hat{f}_n(\xi) \underset{\mathcal{S}}{\to} \hat{f}(\xi)$, and therefore the mapping $\hat{f} : \mathcal{S} \to \mathcal{S}$ is continuous. □

Finally we are prepared to prove the central theorem of Fourier inversion on $\mathcal{S}$.

Theorem 5.11 (Fourier inversion theorem on $\mathcal{S}$). *Suppose that $f \in \mathcal{S}$; then*

$$f(x) = \frac{1}{\sqrt{2\pi}^{n}}\int e^{i\xi\cdot x}\hat{f}(\xi)\,d\xi = \mathcal{F}^{-1}(\hat{f})(x)\ .$$

Proof. Let's first note that this works if $f(x)$ is a Gaussian. For $G(x) = \frac{1}{\sqrt{2\pi}^{n}}e^{-\frac{|x|^2}{2}}$, $\hat{G}(\xi) = \frac{1}{\sqrt{2\pi}^{n}}e^{-\frac{|\xi|^2}{2}}$ by explicit calculation. One interpretation of this fact is that $G(x)$ is an eigenfunction of the Fourier transform, $\mathcal{F}(G) = G$, with eigenvalue 1. Now for the case of general $f(x) \in \mathcal{S}$, use properties given in Lemma (5.7),

$$\begin{aligned}f(0) &= \lim_{\varepsilon\to 0}\int f(x)\frac{1}{\varepsilon^n}G(\frac{x}{\varepsilon})\,dx = \lim_{\varepsilon\to 0}\int f(x)\frac{1}{\varepsilon^n}\sigma_{\frac{1}{\varepsilon}}(G)\,dx \\ &= \lim_{\varepsilon\to 0}\int f(x)\frac{1}{\varepsilon^n}\sigma_{\frac{1}{\varepsilon}}(\hat{G})\,dx = \lim_{\varepsilon\to 0}\int f(x)\widehat{\sigma_{\varepsilon}(G)}\,dx \\ &= \lim_{\varepsilon\to 0}\int \hat{f}(\xi)\sigma_{\varepsilon}(G)(\xi)\,d\xi\ .\end{aligned}$$

We have to verify the last step of this sequence of calculations.

Lemma 5.12. *For $f, g \in \mathcal{S}$,*

$$\int f(x)\overline{\hat{g}(-x)}dx = \int \hat{f}(\xi)\overline{g(\xi)}d\xi\ .$$

Proof of Lemma 5.12.

$$\int f(x)\frac{1}{\sqrt{2\pi}^n}\int e^{-i\xi\cdot x}\overline{g(\xi)}\,d\xi\,dx = \frac{1}{\sqrt{2\pi}^n}\iint f(x)e^{-i\xi\cdot x}\,dx\,\overline{g(\xi)}\,d\xi$$
$$= \int \hat{f}(\xi)\overline{g(\xi)}\,d\xi\ .$$

□

Now apply (5.7) to the result. Since $G(\xi)\big|_{\xi=0} = \frac{1}{\sqrt{2\pi}^n}$,

$$f(0) = \lim_{\varepsilon\to 0}\int \hat{f}(\varepsilon)G(\varepsilon/\xi)\,d\xi = \frac{1}{\sqrt{2\pi}^n}\int \hat{f}(\xi)\,d\xi\ .$$

This recovers $f(0) = \frac{1}{\sqrt{2\pi}^n}\int \hat{f}(\xi)\,d\xi$. To generalize this procedure to cover all $x \in \mathbb{R}^n$ is in fact simple, again using the list of properties in Proposition 5.9:

$$f(x) = (\tau_{-x}f)(0) = \frac{1}{\sqrt{2\pi}^n}\int \mathcal{F}(\tau_{-x}f)(\xi)\,d\xi = \frac{1}{\sqrt{2\pi}^n}\int e^{i\xi\cdot x}\hat{f}(\xi)\,d\xi\ .$$

□

We can finish this circle of ideas with a proof of the $L^2(\mathbb{R}^n)$ Fourier inversion theorem. We have already stated that Schwartz class $\mathcal{S} \subseteq L^2(\mathbb{R}^n)$ is a dense subspace. Lemma 5.12 states that for $f, g \in \mathcal{S}$,

$$\langle f, \mathcal{F}^{-1}g\rangle = \langle \mathcal{F}f, g\rangle\ ,$$

where we also note that

$$(\mathcal{F}^{-1}g)(x) = \hat{g}(-x)\ .$$

Therefore for every $f \in \mathcal{S}$,

$$\begin{aligned}\|f\|_{L^2}^2 &= \langle f, \mathcal{F}^{-1}(\mathcal{F}f)\rangle\\ &= \langle \mathcal{F}(f), \mathcal{F}(f)\rangle\\ &= \|\hat{f}\|_{L^2}^2\ ,\end{aligned}$$

so the Fourier transform acts on the the subspace $\mathcal{S} \subseteq L^2(\mathbb{R}^n)$ isometrically (in the sense of the L^2-norm). Since $\mathcal{S}$ is dense, given an arbitrary $f \in L^2(\mathbb{R}^n)$, there is a sequence $f_n \to f$ in the $L^2(\mathbb{R}^n)$ sense of convergence, with $f_n \in \mathcal{S}$. Then we define

$$\mathcal{F}^{(1)}(f) := \lim_{n\to\infty}\mathcal{F}(f_n)\ .$$

Theorem 5.13 (Functional analysis). *Suppose that an operator* $\mathbf{T}$ *is defined on a dense subspace* $\mathcal{S} \subseteq \mathcal{B}$, *where* $\mathcal{B}$ *is a Banach space, and it is bounded;* $\|\mathbf{T}f\|_{\mathcal{B}} \leq C\|f\|_{\mathcal{B}}$ *for all* $f \in \mathcal{S}$. *Then there exists a unique extension* $\mathbf{T}^{(1)}$ *of* $\mathbf{T}$ *to all of* $\mathcal{B}$, *which is bounded with the same bound.*

Proof. For arbitrary $f \in \mathcal{B}$, consider a sequence $\{f_n\}_{n=1}^{\infty} \to f$ with $\{f_n\}_{n=1}^{\infty} \subseteq \mathcal{S}$. Then $\{f_n\}_{n=1}^{\infty}$ is Cauchy, and so is the sequence $\{\mathbf{T}f_n\}_{n=1}^{\infty}$ because

$$\|\mathbf{T}f_n - \mathbf{T}f_m\|_{\mathcal{B}} \leq C\|f_n - f_m\|_{\mathcal{B}} \to 0 \ .$$

Thus $\{\mathbf{T}f_n\}_{n=1}^{\infty}$ has a limit $g \in \mathcal{B}$, which we define to be $\mathbf{T}^{(1)}f := g$. The extension is clearly linear. It is also well defined, for if $\{h_n\}_{n=1}^{\infty} \subseteq \mathcal{S}$ is another sequence such that $\{h_n\}_{n=1}^{\infty} \to f$, but $Th_n \to g_1$, then

$$\|g - g_1\|_{\mathcal{B}} = \lim_{n\to\infty} \|\mathbf{T}f_n - \mathbf{T}h_n\|_{\mathcal{B}} \leq C \lim_{n\to\infty} \|f_n - h_n\|_{\mathcal{B}} = 0 \ .$$

□

The conclusion that is relevant to the Fourier transform is that $\mathcal{F}$ restricted to $\mathcal{S}$ (such that it has norm $\|\mathcal{F}f\|_{L^2} = \|f\|_{L^2}$ on $\mathcal{S}$) extends uniquely to $\mathcal{F}^{(1)}$ defined on all of $L^2(\mathbb{R}^n)$. Furthermore, this extension is an isometry of $L^2(\mathbb{R}^n)$, as for $\{f_n\}_{n=1}^{\infty} \subseteq \mathcal{S} \subseteq L^2(\mathbb{R}^n)$, with $f_n \to f$:

$$\|f\|_{L^2}^2 = \lim_{n\to\infty} \|f_n\|_{L^2}^2 = \lim_{n\to\infty} \|\hat{f}_n\|_{L^2}^2 = \|\mathcal{F}^{(1)}(f)\|_{L^2}^2 \ .$$

This completes the proof of Theorem 5.4. We will redefine the "extension notation" $\mathcal{F}^{(1)} := \mathcal{F}$ at this time, dropping the extra superscript. The statement is then that $\|f\|_{L^2} = \|\mathcal{F}f\|_{L^2}$ for all $f \in L^2$, which is the desired result.

To finish our remarks we should note that complex isometries $\mathbf{U}$ have the property that $\mathbf{U}^* = \mathbf{U}^{-1}$; namely, they are unitary operators. Let's verify this with the Fourier transform.

Definition 5.14. The adjoint $\mathbf{T}^*$ of an operator $\mathbf{T}$ on a Hilbert space H satisfies $\langle \mathbf{T}^* f, g\rangle = \langle f, \mathbf{T}g\rangle$ for all $f, g \in H$.

Applying this definition of the adjoint to $\mathcal{F}$, take any two $f, g \in L^2$. Then

$$\begin{aligned}\int_{\mathbb{R}^n} g(\xi)\overline{\mathcal{F}(f)(\xi)}\, d\xi &= \int_{\mathbb{R}^n} g(\xi)\overline{\frac{1}{\sqrt{2\pi}^n}\int e^{-i\xi\cdot x} f(x)\, dx}d\xi \\ &= \int_{\mathbb{R}^n} \Big(\int_{\mathbb{R}^n} \frac{1}{\sqrt{2\pi}^n} e^{i\xi\cdot x} g(\xi)\, d\xi\Big)\overline{f(x)}\, dx \\ &= \int_{\mathbb{R}^n} \mathcal{F}^{-1}(g)(x)\overline{(f(x))}\, dx \ ,\end{aligned}$$

which states that $\mathcal{F}^* = \mathcal{F}^{-1}$, as desired.

5.3. Fourier transform of L^1-integrable functions

The standard space of *integrable* functions is $L^1(\mathbb{R}^1)$, which has arisen in our discussions of the heat equation and in particular its interpretation in terms of probability. The space of functions $f \in L^1(\mathbb{R}^n)$ is a Banach space but not a Hilbert space, and in particular there is no way to consistently define an inner product. The theory of the Fourier transform of L^1 functions is subsequently more delicate than its Hilbert space analog. Nonetheless the Fourier transform of a L^1 function is well defined; it has certain properties that will be described below, and the inverse transform is well defined and unique. More subtle details include the manner in which one recovers the function $f(x)$ from its Fourier transform and the characterization of the space $\mathcal{F}L^1(\mathbb{R}^n)$ which is the range of the Fourier transform of L^1. We discuss the first but not the second of these items below. An excellent reference is Dym and McKean [**5**]. For reasons of convenience we will consider the case of one space dimension, namely $L^1(\mathbb{R}^1)$, although all the results below have their direct analogs in $L^1(\mathbb{R}^n)$.

Theorem 5.15. *For $f \in L^1(\mathbb{R}^1)$, define as usual the Fourier transform*

$$\hat{f}(\xi) = \frac{1}{\sqrt{2\pi}} \int e^{-i\xi x} f(y)\, dy \ .$$

The Fourier transform $\hat{f}(\xi)$ of the function $f(x)$ has the following properties.

(i) *Boundedness: $|\hat{f}(\xi)|_{L^\infty} \le \frac{1}{\sqrt{2\pi}} \|f\|_{L^1}$.*

(ii) *Continuity: $\hat{f}(\xi) \in C(\mathbb{R}^1)$.*

(iii) *Limits as $|\xi| \to +\infty$ (Riemann–Lebesgue Lemma):*

$$\lim_{|\xi|\to+\infty} \hat{f}(\xi) = 0 \ .$$

(iv) *Uniqueness: $\hat{f}(\xi) = 0$ if and only if $f(x) = 0$.*

(v) *Convolution algebra:*

$$\widehat{f * g}(\xi) = \sqrt{2\pi}\hat{f}\hat{g}(\xi) \ .$$

Proof. The statement (i) of boundedness follows directly from the expression of the Fourier transform; indeed

$$|\hat{f}(\xi)| = \frac{1}{\sqrt{2\pi}} \Big| \int e^{-i\xi y} f(y)\, dy \Big| \le \frac{1}{\sqrt{2\pi}} \int |f(y)|\, dy \ .$$

To prove the property of continuity (ii), consider the difference

$$\hat{f}(\xi_1) - \hat{f}(\xi_2) = \frac{1}{\sqrt{2\pi}} \int [e^{-i\xi_1 y} - e^{-i\xi_2 y}] f(y)\, dy \ .$$

Given $f \in L^1$, for any $\varepsilon > 0$ we will show that if $|\xi_1 - \xi_2|$ is sufficiently small, then this difference between $f(\xi_1)$ and $f(\xi_2)$ is less than ε in absolute value.

We may write this difference as

$$\hat{f}(\xi_1) - \hat{f}(\xi_2) = \frac{1}{\sqrt{2\pi}} \int_{-R}^{+R} e^{-i\xi_1 y}[1 - e^{-i(\xi_2-\xi_1)y}] f(y)\, dy$$
$$+ \Big(\int_{-\infty}^{-R} + \int_{+R}^{+\infty} \Big) [e^{-i\xi_1 y} - e^{-i\xi_2 y}] f(y)\, dy \ .$$

Since $f \in L^1$, then given $\varepsilon > 0$, there is $R > 0$ such that

$$\Big(\int_{-\infty}^{-R} + \int_{+R}^{+\infty} \Big) |f(y)|\, dy \leq \frac{\varepsilon}{3} \ .$$

Therefore the second term of the RHS is bounded by $2\varepsilon/3$ for R chosen sufficiently large. The first term of the RHS is bounded by

$$\frac{1}{\sqrt{2\pi}} \Big| \int_{-R}^{+R} e^{-i\xi_1 y}[1 - e^{-i(\xi_2-\xi_1)y}] f(y)\, dy \Big| \leq \sup_{|y| \leq R} |[1 - e^{-i(\xi_2-\xi_1)y}]| \int |f(y)|\, dy \ .$$

As long as $|\xi_1 - \xi_2| < \varepsilon \frac{\pi}{6R}$ this term is also bounded by $\varepsilon/3$, and the continuity result follows.

Statement (iii) of the theorem is a result from the theory of the Lebesgue integral. Without going into a full proof, we just note that it is consistent with the expression of $\hat{f}(\xi)$ for $f(x) = \chi_{[a,b]}(x)$, an indicator function of an interval:

$$\frac{1}{\sqrt{2\pi}} \int e^{-i\xi y} \chi_{[a,b]}(y)\, dy = \int_a^b e^{-i\xi y}\, dy = \frac{e^{-i\xi b} - e^{-i\xi a}}{-i\xi} \ ,$$

which has this decay property as $|\xi| \to +\infty$.

Statement (v) concerns the properties of convolution, which are that

$$\mathcal{F}(f * g)(\xi) = \frac{1}{\sqrt{2\pi}} \int e^{-i\xi y} \int f(y - y') g(y')\, dy' dy$$
$$= \frac{1}{\sqrt{2\pi}} \iint e^{-i\xi(y-y')} f(y - y')\, dy \ \ e^{-i\xi y'} g(y')\, dy' = \sqrt{2\pi} \hat{f}(\xi) \hat{g}(\xi) \ .$$

Fubini's theorem is invoked to justify the exchange of integrations when both $f, g \in L^1$.

Finally, for statement (iv) consider $f \in L^1$ while $g \in \mathcal{S}$ is an arbitrary Schwartz class test function. Then

$$\int f(x) g(x)\, dx = \int f(x) \frac{1}{\sqrt{2\pi}} \int e^{-i\xi x} \hat{g}(-\xi)\, d\xi dx$$
$$= \int \hat{f}(\xi) \hat{g}(-\xi)\, d\xi \ ,$$

again using Fubini's theorem to interchange integrals. If we have $\hat{f}(\xi) = 0$, then

$$\int f(x)g(x)\,dx = 0$$

for all $g \in \mathcal{S}$, implying that $f(x)$ itself must vanish identically. □

As remarked above, the convergence of the inverse Fourier transform of a function in L^1 is less straightforward than in the case of L^2. In this direction we describe two results. It would seem that the simplest manner to invert the Fourier transform of $\hat{f}(\xi)$ would be as the limit of partial inverses

$$\lim_{R\to+\infty} f_R(x) := \lim_{R\to+\infty} \frac{1}{\sqrt{2\pi}} \int_{-R}^{+R} e^{i\xi x} \hat{f}(\xi)\,d\xi \ .$$

However, this does not converge well in general; in fact for the analogous question for convergence of Fourier series, there are counterexamples to convergence in what are called lacunary sequences. An alternate limiting process involves taking an average of such partial inverses. In particular we take limits of the T-averages of the above expression,

$$\frac{1}{T}\int_0^T \Big(\frac{1}{\sqrt{2\pi}} \int_{-R}^{+R} e^{i\xi x} \hat{f}(\xi)\,d\xi\Big)\,dR$$

as $T \to +\infty$, which turns out to have substantially better convergence properties.

Theorem 5.16 (Cesaro integrability). *Given $f \in L^1(\mathbb{R}^1)$,*

$$\lim_{T\to+\infty} \frac{1}{T}\int_0^T \Big(\frac{1}{\sqrt{2\pi}} \int_{-R}^{+R} e^{i\xi x} \hat{f}(\xi)\,d\xi\Big)\,dR = f(x) \ , \tag{5.3}$$

where the convergence of the functions on the LHS to the RHS is taken with respect to the L^1-norm.

Proof. First of all we can describe the integrals over $[-R,+R]$ in terms of a convolution

$$\begin{aligned}
\sqrt{2\pi}\int_{-R}^{+R} e^{i\xi x}\hat{f}(\xi)\,d\xi &= \int_{-R}^{+R}\int_{-\infty}^{+\infty} e^{i\xi(x-y)} f(y)\,dyd\xi \\
&= \int_{-\infty}^{+\infty}\Big(\int_{-R}^{+R} e^{i\xi(x-y)}\,d\xi\Big)\,f(y)\,dy \\
&= \int_{-\infty}^{+\infty}\Big(\frac{e^{iR(x-y)} - e^{-iR(x-y)}}{i(x-y)}\Big)\,f(y)\,dy \ .
\end{aligned}$$

Using this expression, we now average over $0 \le R \le T$,

$$\begin{aligned}
&\frac{1}{T}\int_0^T \Big(\frac{1}{\sqrt{2\pi}}\int_{-R}^{+R} e^{i\xi x}\hat{f}(\xi)\,d\xi\Big)\,dR \\
&\quad = \frac{1}{T}\int_{-\infty}^{+\infty}\int_0^T \frac{1}{\sqrt{2\pi}}\Big(\frac{e^{iR(x-y)} - e^{-iR(x-y)}}{i(x-y)}\Big)\,dR\, f(y)\,dy \\
&\quad = \int_{-\infty}^{+\infty}\frac{1}{T}\Big(\frac{\sin^2(T(x-y))}{\sqrt{2\pi}(x-y)^2}\Big)\, f(y)\,dy \ ,
\end{aligned}$$

which is also a convolution of $f(x)$, with the kernel function

$$F_T(x-y) = \frac{1}{T}\Big(\frac{\sin^2(T(x-y))}{\sqrt{2\pi}(x-y)^2}\Big) \ .$$

The kernel $F_T(x)$ is the analog of the Féjer kernel in the proof of Cesaro summability of Fourier series. It is nonnegative and its integral equals one. Therefore the task of the proof of convergence is to show that for arbitrary $\varepsilon > 0$, the difference

$$\begin{aligned}
\int F_T(x-y)f(y)\,dy - f(x) &= \int F_T(y)[f(x-y) - f(x)]\,dy \\
&= \int_{-\infty}^{+\infty}\frac{1}{T}\frac{\sin^2(Ty)}{y^2}[f(x-y) - f(y)]\,dy
\end{aligned}$$

can be made smaller than ε when measured in L^1. The L^1-norm of this difference is

$$\begin{aligned}
&\int_{-\infty}^{+\infty}\Big|\int_{-\infty}^{+\infty} F_T(y)[f(x-y) - f(x)]\,dy\Big|\,dx \\
&\le \int_{-\infty}^{+\infty}\int_{-\infty}^{+\infty}\frac{1}{T}\frac{\sin^2(Ty)}{y^2}\big|[f(x-y) - f(x)]\big|\,dx\,dy \\
&\le \int_{-\delta}^{+\delta}\int_{-\infty}^{+\infty}\frac{1}{T}\frac{\sin^2(Ty)}{y^2}\big|[f(x-y) - f(x)]\big|\,dx\,dy \\
&\quad + \Big(\int_{-\infty}^{-\delta} + \int_{+\delta}^{+\infty}\Big)\int_{-\infty}^{+\infty}\frac{1}{T}\frac{\sin^2(Ty)}{y^2}\big|[f(x-y) - f(x)]\big|\,dx\,dy \ . \qquad (5.4)
\end{aligned}$$

Since translation by $y \in \mathbb{R}^1$ is continuous in the L^1-norm, for y restricted to $[-\delta, +\delta]$ and for $\delta > 0$ sufficiently small,

$$\begin{aligned}
&\int_{-\delta}^{+\delta}\int_{-\infty}^{+\infty}\frac{1}{T}\frac{\sin^2(Ty)}{y^2}\big|[f(x-y) - f(x)]\big|\,dx\,dy \\
&\quad \le \|f(\cdot - y) - f(\cdot)\|_{L^1}\int_{-\delta}^{+\delta} F_T(y)\,dy < \frac{\varepsilon}{2} \ .
\end{aligned}$$

The remaining term of the RHS of (5.4) is also bounded by $\varepsilon/2$ for T sufficiently large, namely

$$\begin{aligned}
&\Big(\int_{-\infty}^{-\delta} + \int_{+\delta}^{+\infty}\Big) \int_{-\infty}^{+\infty} \frac{1}{T}\frac{\sin^2(Ty)}{y^2} \big|[f(x-y) - f(x)]\big| \, dx\, dy \\
&\leq 2\|f\|_{L^1} \Big(\int_{-\infty}^{-\delta} + \int_{+\delta}^{+\infty}\Big) \frac{1}{T}\frac{\sin^2(Ty)}{y^2}\, dy \\
&\leq 2\|f\|_{L^1} \Big(\int_{-\infty}^{-T\delta} + \int_{+T\delta}^{+\infty}\Big) \frac{\sin^2(y')}{(y')^2}\, dy' \ .
\end{aligned}$$

Since the function $\sin^2(y')/(y')^2$ is integrable, for $\delta > 0$ fixed, one can take T large so that this latter term is also bounded above by $\varepsilon/2$. □

A second setting is related to solutions of the heat equation, which also can be considered as a smoothing or an averaging process.

Theorem 5.17. *Suppose that $f \in L^1(\mathbb{R}^1)$ and that $u(t,x)$ is the solution of the heat equation with initial data $f(x)$, given by $u(t,x) = H(t,\cdot) * f(x)$. Then*

$$f(x) = \lim_{t\to 0+} u(t,x) = \lim_{t \to 0+} \frac{1}{\sqrt{2\pi}} \int e^{i\xi x} e^{-\frac{1}{2}|\xi|^2 t} \hat{f}(\xi)\, d\xi \ ,$$

where again the limit is in the sense of convergence in the $L^1(\mathbb{R}^1)$-norm.

Proof. The heat kernel has Fourier transform $\widehat{H}(t,\xi) = e^{-\frac{1}{2}|\xi|^2 t}$ of course, and for $t > 0$, it is a member of Schwartz class. From Theorem 5.15(ii), we know that $\hat{f}(\xi)$ is bounded; thus the product satisfies $e^{-\frac{1}{2}|\xi|^2 t}\hat{f}(\xi) \in L^1(\mathbb{R}^1)$. Hence it has a well-defined inverse Fourier transform, which, again by the Fubini theorem, satisfies

$$\mathcal{F}(e^{-\frac{1}{2}|\xi|^2 t}\hat{f})(x) = (H(t,\cdot) * f)(x) \ .$$

It therefore suffices to study the limit as $t \to 0+$ of the quantity

$$\begin{aligned}
\|H(t,\cdot) * f - f\|_{L^1} &= \int \Big| \int H(t, x-y)[f(y) - f(x)]\, dy \Big|\, dx \\
&\leq \int \int_{-\delta}^{+\delta} \big|H(t,y)[f(x-y) - f(x)]\big|\, dy\, dx \\
&\quad + \int \Big(\int_{-\infty}^{-\delta} + \int_{+\delta}^{+\infty}\Big) \big|H(t,y)[f(x-y) - f(x)]\big|\, dy\, dx \ .
\end{aligned}$$

Consider an arbitrary $\varepsilon > 0$. Use the fact that translation by $y \in \mathbb{R}^1$ is continuous in the L^1-norm so that for $\delta > 0$ sufficiently small,

$$\int_{-\delta}^{+\delta} \int \big|H(t,y)[f(x-y) - f(x)]\big|\, dx\, dy \leq \sup_{|y|\leq\delta} \|f(\cdot - y) - f(\cdot)\|_{L^1} \leq \frac{\varepsilon}{2} \ .$$

The remaining term of the difference is bounded by

$$\int \Big(\int_{-\infty}^{-\delta} + \int_{+\delta}^{+\infty} \Big) \big| H(t,y)[f(x-y) - f(x)] \big| \, dy \, dx$$
$$\le 2\|f\|_{L^1} \Big(\int_{-\infty}^{-\delta} + \int_{+\delta}^{+\infty} \Big) H(t,y) \, dy \ .$$

As long as $0 < t$ is taken sufficiently small, this quantity can also be shown to be less than $\varepsilon/2$, as in the estimate (3.8) of the heat kernel. □

The results of this subsection are related in spirit to Theorem 3.7 which concerns the pointwise behavior of convolutions with the heat kernel as $t \to 0+$. It is the statement that at points x_0 at which $f(x)$ is continuous, $\lim_{t\to 0+} u(t, x_0) = f(x_0)$. A similar result holds for the Cesaro integrals, with a similar proof. Namely, if $f(x)$ is continuous at x_0, then

$$\lim_{T\to+\infty} \int F_T(x_0 - y) f(y) \, dy = f(x_0) \ .$$

This pointwise limit at points of continuity of $f(x)$ is the subject of Exercise 5.8.

Exercises: Chapter 5

Exercise 5.1. Show that the Fourier transform preserves angles between vectors. For $f, g \in L^2$,

$$Re\langle f, g\rangle = \|f\|_{L^2} \|g\|_{L^2} \cos\theta$$

and

$$Re\langle \hat{f}, \hat{g}\rangle = \|\hat{f}\|_{L^2} \|\hat{g}\|_{L^2} \cos\varphi \ .$$

Show that $\theta = \varphi$.

Exercise 5.2. The $L^p(\Omega)$-norm for functions on a domain $\Omega \subseteq \mathbb{R}^n$ is defined as

$$\|f\|_{L^p} = \Big(\int_\Omega |f(x)|^p \, dx \Big)^{1/p}$$

for $1 \le p < +\infty$. The Banach space $L^\infty(\Omega)$ is defined for $p = +\infty$ by the norm

$$\|f\|_{L^\infty} = \sup_{x\in\Omega} |f(x)| \ .$$

Because of Lebesgue measure, this supremum is taken modulo sets of zero measure (the *essential supremum*). The Hölder inequality states that

$$\Big| \int_\Omega f(x) g(x) \, dx \Big| \le \|f\|_{L^p} \|g\|_{L^{p'}} \ ,$$

where $\frac{1}{p} + \frac{1}{p'} = 1$ are dual indices. A special case for $p = p' = 2$ is the Cauchy–Schwarz inequality. Give a proof of the Hölder inequality for the range of p, p' given above.

Exercise 5.3. This problem concerns the cases of domains $\Omega = \mathbb{R}^n$ and bounded domains $\Omega \subseteq \mathbb{R}^n$.

(1) In the case of $\Omega = \mathbb{R}^n$, for which indices p and q does
$$L^p(\mathbb{R}^n) \subseteq L^q(\mathbb{R}^n)$$
hold?

(2) In the case of bounded Ω, use the Hölder inequality to show that
$$L^p(\Omega) \subseteq L^q(\Omega)$$
for $q \le p$.

(3) In the case of bounded Ω, is it true that
$$L^\infty(\Omega) = \bigcap_{1 \le p < +\infty} L^p(\Omega) \ ?$$

Exercise 5.4. Prove part (i) of the technical Lemma 5.7 on Schwartz class functions. Namely, show that for $g \in \mathcal{S}(\mathbb{R}^n)$ such that $g(0) = 1$, then for all $f \in \mathcal{S}(\mathbb{R}^n)$,
$$\mathcal{S} - \lim_{\varepsilon \to 0} \big(g(\varepsilon x) f(x)\big) = f(x) \ .$$

Exercise 5.5. Give examples of nonanalytic C^∞ functions on $\mathbb{R}^n$. *Hint:* See Project 3.12 of Chapter 3.

Exercise 5.6. Prove that Schwartz class $\mathcal{S}$ is a dense subspace of $L^2(\mathbb{R}^n)$. *Hints:* Show that the subspace of compactly supported functions is dense in $L^2(\mathbb{R}^n)$. Then show that every compactly supported function in $L^2(\mathbb{R}^n)$ can be approximated in the L^2-norm by a sequence of functions in $C_0^\infty \subseteq \mathcal{S}$.

Exercise 5.7. We have described the Fourier transform $\mathcal{F}$ as a unitary operator on $L^2(\mathbb{R}^n)$. What are the eigenvalues and eigenfunctions $(\lambda_k, \psi_k(x))$ of this operator? For simplicity you may consider only the case $n = 1$. Note: this exercise is related to Exercise 2.6 of Chapter 2.

Exercise 5.8. Suppose that $f \in L^1(\mathbb{R}^1)$ and that $f(x)$ is continuous at the point $x = x_0$. Prove that for large T, the limit satisfies
$$\lim_{T \to +\infty} \frac{1}{T} \int_0^T \Big(\frac{1}{\sqrt{2\pi}} \int_{-R}^{+R} e^{i\xi x_0} \hat{f}(\xi) \, d\xi \Big) \, dR = f(x_0) \ .$$

Chapter 6

Wave equations on $\mathbb{R}^n$

Numerical simulation of a merger of two black holes generating gravitational waves, similar to those observed by the LIGO detectors. Image credit: SXS, the Simulating eXtreme Spacetimes (SXS) project (`http://www.black-holes.org`).

Solutions of the wave equation describe the propagation of light, sound waves in a gas or fluid, gravitational waves in the interstellar vacuum, and many other phenomena. It is one of my favorite equations. Posed in $\mathbb{R}^1_t \times \mathbb{R}^n_x$, the initial value problem (or *Cauchy problem*) for the equation looks very similar to (2.13) of Chapter 2:

$$\partial_t^2 u - \Delta u = 0, \qquad x \in \mathbb{R}^n, t \in \mathbb{R}^1 \ , \tag{6.1}$$

with initial or Cauchy data for $u(x,t)$ given by

$$u(0,x) = f(x) \ , \qquad \partial_t u(0,x) = g(x) \ .$$

The given initial data $u(0,x) = f(x)$ is often referred to as the initial position or displacement of the field $u(t,x)$, while the data for $\partial_t u(0,x) = g(x)$ is called the initial velocity, or momentum, as it specifies the time derivative of the field $u(t,x)$ at the initial moment.

6.1. Wave propagator by Fourier synthesis

Assuming first that $f(x), g(x) \in \mathcal{S}$, that is, it is in Schwartz class, we may take the Fourier transform of the equation to obtain

$$\partial_t^2 \hat{u}(t,\xi) + |\xi|^2 \hat{u}(t,\xi) = 0 \ . \tag{6.2}$$

Solutions of this second-order ODE are composed of linear combinations of $e^{\pm i|\xi|t}$; taking into account that $\hat{u}(0,\xi) = \hat{f}(\xi)$ and $\partial_t \hat{u}(0,\xi) = \hat{g}(\xi)$, we derive the expression

$$\hat{u}(t,\xi) = \cos(|\xi|t)\hat{f}(\xi) + \frac{\sin(|\xi|t)}{|\xi|}\hat{g}(\xi) \ . \tag{6.3}$$

The inverse Fourier transform gives the solution

$$u(t,x) = \frac{1}{\sqrt{2\pi}^n} \int e^{i\xi\cdot x} \left(\cos(|\xi|t)\hat{f}(\xi) + \frac{\sin(|\xi|t)}{|\xi|}\hat{g}(\xi) \right) d\xi \ . \tag{6.4}$$

Observe from (6.3) that $\hat{u}(t,\xi) \in \mathcal{S}(\mathbb{R}^n_x)$ for each time t, and therefore our solution given in (6.4) is in Schwartz class as well. However, this high level of smoothness is not at all necessary, and one notes that expression (6.4) makes sense whenever $\hat{f}(\xi), \frac{\hat{g}(\xi)}{|\xi|} \in L^2(\mathbb{R}^n_x)$, in which case it may not possess derivatives in the classical sense.

Theorem 6.1. *For $f, g \in \mathcal{S}$, expression* (6.4) *gives a solution of the wave equation $u(t,x) \in \mathcal{S}(\mathbb{R}^n_x)$ for each time $t \in \mathbb{R}$. For $\hat{f}(\xi), \frac{\hat{g}(\xi)}{|\xi|} \in L^2(\mathbb{R}^n_x)$,* (6.4) *gives a weak solution to the wave equation in the sense that* (6.2) *is satisfied.*

One basic property satisfied by solutions of the wave equation is the principle of "energy" conservation under time evolution. The energy of a solution is defined to be the expression

$$E(u) = \frac{1}{2} \int (\partial_t u(t,x))^2 + |\nabla_x u(t,x)|^2 \, dx \ . \tag{6.5}$$

Theorem 6.2. *For $(f,g) \in H^1(\mathbb{R}^n_x) \times L^2(\mathbb{R}^n_x)$, the energy* (6.5) *is conserved for solutions of the wave equation* (6.1).

Proof. Using expression (6.3) for the Fourier transform of the solution, we give a similar formula for the vector $(\partial_x u(t,x), \partial_t u(t,x))^T$:

$$\begin{pmatrix} \partial_x u(t,x) \\ \partial_t u(t,x) \end{pmatrix} = \frac{1}{\sqrt{2\pi}^n} \int e^{i\xi\cdot x} \begin{pmatrix} \cos(|\xi|t) & \frac{i\xi}{|\xi|}\sin(|\xi|t) \\ \frac{i\xi}{|\xi|}\sin(|\xi|t) & \cos(|\xi|t) \end{pmatrix} \begin{pmatrix} i\xi\hat{f}(\xi) \\ \hat{g}(\xi) \end{pmatrix} d\xi \ .$$

By inspection this is a well-defined vector valued function in $[L^2(\mathbb{R}^n)]^{n+1}$ as long as $(\nabla_x f(x), g(x))^T \in [L^2(\mathbb{R}^n)]^{n+1}$, which it is by hypothesis. By the Plancherel identity of Theorem 5.4,

$$\|\partial_t u(x,t)\|_{L^2}^2 + \|\partial_x u(x,t)\|_{L^2}^2 = \| \begin{pmatrix} \cos(|\xi|t) & \frac{i\xi}{|\xi|}\sin(|\xi|t) \\ \frac{i\xi}{|\xi|}\sin(|\xi|t) & \cos(|\xi|t) \end{pmatrix} \begin{pmatrix} i\xi\hat{f}(\xi) \\ \hat{g}(\xi) \end{pmatrix} \|_{L^2}^2$$
$$= \| \begin{pmatrix} i\xi\hat{f}(\xi) \\ \hat{g}(\xi) \end{pmatrix} \|_{L^2}^2 = \|\nabla_x f\|_{L^2}^2 + \|g\|_{L^2}^2 \ .$$

The second to last equality holds because the 2×2 matrix featured in this calculation is unitary. □

An alternative proof works in the case that we also have $\partial_t^2 u, \Delta u \in L^2$; then for solutions of (6.1),

$$\frac{d}{dt}E(u) = \frac{1}{2}\int_{\mathbb{R}^n} 2\partial_t u \partial_t^2 u + 2\partial_x u \partial_t \partial_x u \, dx$$
$$= \int_{\mathbb{R}^n} \partial_t u \left(\partial_t^2 u - \Delta u\right) \, dx = 0 \ .$$

6.2. Lorentz transformations

Just as the Laplace operator $\Delta = \sum_{j=1}^n \partial_{x_j}^2$ is invariant under translations $x' = x + c$ and rotations $x' = Rx$, where $R^T = R^{-1}$, as we have seen in Chapter 1, Exercise 1.3, the wave operator or d'Alembertian

$$\Box = \partial_t^2 - \Delta = \partial_t^2 - \sum_{j=1}^n \partial_{x_j}^2$$

is invariant under a group of transformations of the space-time $\mathbb{R}_t^1 \times \mathbb{R}_x^n$ known as the Lorentz group. Elements of this transformation group are generated by the same spatial rotations of $\mathbb{R}_x^n$ as above and *hyperbolic rotations* which involve time as well as space. Define a hyperbolic rotation in the (t, x_1) coordinate plane in $\mathbb{R}^2 \subseteq \mathbb{R}^{n+1}$ by

$$\begin{pmatrix} t' \\ x'_1 \\ \vdots \\ x'_n \end{pmatrix} = \begin{pmatrix} \cosh(\psi) & \sinh(\psi) & \dots & 0 \\ \sinh(\psi) & \cosh(\psi) & & \\ \vdots & & I_{n-1\times n-1} & \\ 0 & \dots & & \end{pmatrix} \begin{pmatrix} t \\ x_1 \\ \vdots \\ x_n \end{pmatrix} = H(\psi) \begin{pmatrix} t \\ x_1 \\ \vdots \\ x_n \end{pmatrix} .$$

One calculates that $\det(H(\psi)) = \cosh(\psi)^2 - \sinh(\psi)^2 = 1$ and that $H^{-1}(\psi) = H(-\psi)$. Vector fields in the (t, x_1) coordinate plane transform as follows:

$$\begin{pmatrix} \partial_t \\ \partial_{x_1} \end{pmatrix} = \begin{pmatrix} \cosh(\psi) & \sinh(\psi) \\ \sinh(\psi) & \cosh(\psi) \end{pmatrix} \begin{pmatrix} \partial_{t'} \\ \partial_{x'_1} \end{pmatrix} = H_{2\times 2}(\psi) \begin{pmatrix} \partial_{t'} \\ \partial_{x'_1} \end{pmatrix} , \tag{6.6}$$

while all other $\partial_{x_j} = \partial_{x'_j}$ are invariant. The Lorentz group of transformations on the space-time $\mathbb{R}^1_t \times \mathbb{R}^n_x$ is generated by all spatial rotations R and by the hyperbolic rotations $H(\psi)$. It is a Lie group known as $SO(1,n)$. The set of space-time translations $(t', x') = (t+b, x+c)$ leave the wave equation invariant, and so do the elements of the Lorentz group.

Proposition 6.3. *The Lorentz transformations leave the d'Alembertian operator invariant.*

Proof. Since the group generators involving translations obviously leave the d'Alembertian invariant, and all spatial rotations R just involve the Laplacian, it suffices to check invariance under the hyperbolic rotations $H(\psi)$. Write the d'Alembertian operator using matrix notation:

$$\Box u = (\partial_t^2 - \Delta)u = \begin{pmatrix} \partial_t \\ \partial_x \end{pmatrix}^T \begin{pmatrix} 1 & 0 \\ 0 & -I_{n\times n} \end{pmatrix} \begin{pmatrix} \partial_t \\ \partial_x \end{pmatrix} u \ .$$

Use the transformation rule (6.6) for vector fields to check the invariance of the d'Alembertian:

$$\begin{pmatrix} \partial_{t'} \\ \partial_{x'} \end{pmatrix}^T \begin{pmatrix} H^T_{2\times 2} & 0 \\ 0 & I_{(n-1)\times(n-1)} \end{pmatrix} \begin{pmatrix} 1 & 0 \\ 0 & -I_{n\times n} \end{pmatrix} \begin{pmatrix} H_{2\times 2} & 0 \\ 0 & I_{(n-1)\times(n-1)} \end{pmatrix} \begin{pmatrix} \partial_{t'} \\ \partial_{x'} \end{pmatrix} .$$

The only nontrivial part of this matrix calculation is the upper left hand 2×2 block,

$$H^T_{2\times 2} \begin{pmatrix} 1 & 0 \\ 0 & -1 \end{pmatrix} H_{2\times 2} = \begin{pmatrix} \cosh(\psi) & \sinh(\psi) \\ \sinh(\psi) & \cosh(\psi) \end{pmatrix} \begin{pmatrix} 1 & 0 \\ 0 & -1 \end{pmatrix} \begin{pmatrix} \cosh(\psi) & \sinh(\psi) \\ \sinh(\psi) & \cosh(\psi) \end{pmatrix}$$

$$= \begin{pmatrix} \cosh^2(\psi) - \sinh^2(\psi) & 0 \\ 0 & \sinh^2(\psi) - \cosh^2(\psi) \end{pmatrix} = \begin{pmatrix} 1 & 0 \\ 0 & -1 \end{pmatrix} ,$$

which is precisely to say that the d'Alembertian $\Box$ is invariant under $H(\psi)$. This is the stated result. $\Box$

The effects of the Lorentz transformation $H(\psi)$ on the time axis $\{x = 0\}$, exhibited in Figure 1, are given by

$$x'_1 = \sinh(\psi)t \ , \qquad t' = \cosh(\psi)t \ ;$$

therefore their ratio

$$\frac{x'_1}{t'} = \tanh(\psi) := v$$

gives the velocity of the new frame of reference with respect to the old. We note that $|v| = |\tanh(\psi)| < 1$. In terms of v, the hyperbolic rotation can be written in a more familiar form:

$$H = \begin{pmatrix} \frac{1}{\sqrt{1-|v|^2}} & \frac{-v}{\sqrt{1-|v|^2}} \\ \frac{-v}{\sqrt{1-|v|^2}} & \frac{1}{\sqrt{1-|v|^2}} \end{pmatrix} .$$

The coordinate plane $\{(0, 0, x_2, \ldots, x_n)\}$ is invariant under $H(\psi)$, but the x_1-axis $\{(t, x_2, \ldots, x_n) = 0\}$ is moved,

$$x'_1 = \cosh(\psi) x_1 \ , \qquad t' = \sinh(\psi) x_1 \ ,$$

so that the $\{t = 0\}$ coordinate plane is tilted in space-time as well, at slope

$$\frac{x'_1}{t'} = \coth(\psi) \ .$$

A diagram of the new space-time coordinates under a hyperbolic rotation is as follows.

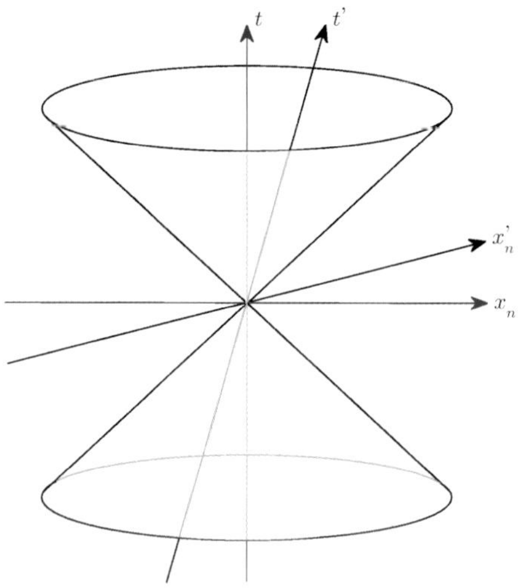

Figure 1. This is the image of the x_1- and t-axes and the light cone under a Lorentz transformation consisting of a hyperbolic rotation of the (t, x_1) coordinates.

The set that remains invariant under the transformations of the Lorentz group is the light cone $LC = \{t^2 - |x|^2 = 0\}$. This fact is checked as follows: take $(t, x) \in LC$, then

$$\begin{pmatrix} t \\ x \end{pmatrix}^T \begin{pmatrix} 1 & 0 \\ 0 & -I_{n\times n} \end{pmatrix} \begin{pmatrix} t \\ x \end{pmatrix} = 0 \ .$$

As we have checked above,

$$L^T \begin{pmatrix} 1 & 0 \\ 0 & -I_{n\times n} \end{pmatrix} L = \begin{pmatrix} 1 & 0 \\ 0 & -I_{n\times n} \end{pmatrix} \ ;$$

therefore under a Lorentz transformation $L : (t, x) \mapsto (t', x')$, we have

$$\begin{pmatrix} t' \\ x' \end{pmatrix}^T \begin{pmatrix} 1 & 0 \\ 0 & -I_{n\times n} \end{pmatrix} \begin{pmatrix} t' \\ x' \end{pmatrix} = \begin{pmatrix} t \\ x \end{pmatrix}^T L^T \begin{pmatrix} 1 & 0 \\ 0 & -I_{n\times n} \end{pmatrix} L \begin{pmatrix} t \\ x \end{pmatrix} = 0 \ .$$

The geometric interpretation of these properties has to do with the matrix

$$g := \begin{pmatrix} 1 & 0 \\ 0 & -I_{n\times n} \end{pmatrix}$$

that defines the Minkowski metric

$$ds^2 = (dt, dx_1, \ldots, dx_n)^T g(dt, dx_1, \ldots, dx_n)$$

on the space-time $\mathbb{R}^1_t \times \mathbb{R}^n_x$. What we have just shown is that this metric is invariant under the Lorentz group.

The invariance of the wave equation under the Lorentz group of transformations was part of a paradox that physics faced towards the end of the nineteenth century. As we have seen in (1.1), Maxwell's equations are intimately tied to the wave equation. The equations of classical mechanics are invariant under the Galilean transformation group, that is, the transformation group consisting of rotations and translations of space along with constant velocity translations. And the theory of electricity and magnetism is invariant under the Lorentz group. The fact is that these are incompatible. Compatibility was restored in 1905 when Einstein introduced the special theory of relativity. However, this was a revolutionary change in our perception of the universe, for which many intuitive ideas about space-time had to be modified. One of these we have seen above; the action of a hyperbolic rotation transforms the plane $\{t = 0\}$ of spatial coordinates as well as the time axis. Since we think of the spatial coordinate plane as being the "present" state of the universe, it is a new idea that there is no universally valid instant that can globally be considered to be the present, and that the sense of the present is relative to the frame of reference of the observer. The concept of simultaneity has to be sacrificed. In terms of the wave equation, any one of the hypersurfaces $\{t' = 0\}$ can be used as a hypersurface on which to pose initial data for the wave equation.

6.3. Method of spherical means

There is another method for representing the solution of the wave equation in n space dimensions (6.1), based on the spherical means that we have encountered in Chapter 4, formula (4.13). Recall the elementary solution method in the case of spatial dimension $n = 1$ via the d'Alembert formula

$$u(t,x) = \frac{1}{2}(f(x+t) + f(x-t)) + \frac{1}{2}\int_{x-t}^{x+t} g(y)\, dy \ .$$

The goal of this section is to produce similar expressions for the solution of the wave equation in the case of higher space dimensions.

Definition 6.4. Given a function $h(x) \in C(\mathbb{R}^n)$, its *spherical mean* centered about the point $x \in \mathbb{R}^n$ is

$$M(h)(x,r) := \frac{1}{\omega_n r^{n-1}} \int_{|x-y|=r} h(y)\, dS_y \ , \tag{6.7}$$

the average of $h(x)$ over a sphere $\mathbb{S}^{n-1}$ centered at x of radius r.

Recall that $\omega_n r^{n-1}$ is the surface area of the sphere $\mathbb{S}^{n-1}$ of radius r in n dimensions. When $n = 1$, the spherical mean of a function $f(x)$ is $M(f)(x,r) = \frac{1}{2}(f(x+r) + f(x-r))$, reminding one of the first term of the d'Alembert formula. Changing variables in the integral (6.7), $y \mapsto x + r\xi$, $\xi \in S_1(0)$, there is another useful expression for the spherical mean:

$$M(h)(x,r) = \frac{1}{\omega_n} \int_{|\xi|=1} h(x+r\xi)\, dS_\xi \ . \tag{6.8}$$

Expression (6.8) is defined for $r \geq 0$, but it is clear from this second expression that $M(h)(x,r)$ is an even function of r:

$$M(h)(x,-r) = M(h)(x,r) \ .$$

Lemma 6.5 (Darboux equation). *Given $h \in C^2(\mathbb{R}^n)$,*

$$\Delta_x M(h)(x,r) = \left(\partial_r^2 + \frac{n-1}{r}\partial_r\right) M(h)(x,r) \ ,$$

a formula that relates the Laplace operator of $M(h)$ in the n-dimensional x-variables to an operator involving only one-dimensional r derivatives.

Proof. The Darboux equation follows from a calculation using multivariate calculus. First take one derivative:

$$\begin{aligned}
\partial_r M(h)(x,r) &= \frac{1}{\omega_n} \int_{|\xi|=1} \partial_r h(x+r\xi)\, dS_\xi \\
&= \frac{1}{\omega_n} \int_{|\xi|=1} \sum_{j=1}^{n} \partial_{x_j} h(x+r\xi)\xi_j\, dS_\xi \ .
\end{aligned}$$

We notice that $\sum_{j=1}^n \partial_{x_j} h(x+r\xi)\xi_j = \nabla h \cdot N$, the outwards normal derivative of h on the unit sphere. Continue this line of calculation using

Green's theorem:

$$\begin{aligned}\frac{1}{\omega_n}\int_{|\xi|=1}\sum_{j=1}^{n}\partial_{x_j}h(x+r\xi)\xi_j\,dS_\xi &= \frac{1}{\omega_n}\int_{|\xi|<1} r\Delta_x h(x+r\xi)\,d\xi \\ &= \frac{1}{\omega_n r^{n-1}}\Delta_x\int_{|x-y|<r} h(y)\,dy \\ &= \frac{1}{\omega_n r^{n-1}}\Delta_x\Big(\int_0^r\int_{|x-y|=\rho} h(y)\,dS_y\,d\rho\Big) \\ &= \frac{1}{r^{n-1}}\Delta_x\Big(\int_0^r \rho^{n-1}M(h)(x,\rho)\,d\rho\Big)\ .\end{aligned}$$

The second step is to take a second derivative after multiplying through by r^{n-1}:

$$\begin{aligned}\partial_r\big(r^{n-1}\partial_r M(h)(x,r)\big) &= \partial_r\Big(\Delta_x\int_0^r \rho^{n-1}M(h)(x,\rho)\,d\rho\Big) \\ &= \Delta_x\big(r^{n-1}M(h)(x,r)\big)\ .\end{aligned}$$

Therefore

$$\begin{aligned}\Delta_x M(h)(x,r) &= \frac{1}{r^{n-1}}\partial_r\big(r^{n-1}\partial_r M(h)(x,r)\big) \\ &= \big(\partial_r^2+\frac{n-1}{r}\partial_r\big)M(h)(x,r)\ ,\end{aligned}\tag{6.9}$$

which is the result of the lemma. □

Proposition 6.6. (i) *For $h(x)\in C(\mathbb{R}^n)$, the value of $h(x)$ at any $x\in\mathbb{R}^n$ can be recovered from its spherical mean:*

$$h(x)=\lim_{r\to 0}M(h)(x,r)=M(h)(x,0)\ .$$

(ii) *Additionally, for $h(x)\in C^2(\mathbb{R}^n)$,*

$$\partial_r M(h)(x,0)=\lim_{r\to 0}\frac{r}{\omega_n}\int_{|\xi|<1}h(x+r\xi)\,d\xi=0\ .$$

Formulae involving spherical means can be used to give an expression for the solution of the wave equation. Indeed, suppose that $u(t,x)$ is a solution of the Cauchy problem for the wave equation (6.1). Then its spherical mean $M(u)(t,x,r)$ can be defined from (6.7), and it satisfies an auxiliary equation in the reduced space-time variables $(t,r)\in\mathbb{R}^2$. Specifically, define

$$M(u)(t,x,r)=\frac{1}{\omega_n}\int_{|\xi|=1}u(t,x+r\xi)\,dS_\xi\ .$$

Taking time derivatives and using the fact that $u(t,x)$ is a solution of (6.1), one obtains

$$\begin{aligned}\partial_t^2 M(u)(t,x,r) &= \frac{1}{\omega_n}\int_{|\xi|=1} \partial_t^2 u(t,x+r\xi)\,dS_\xi \\ &= \frac{1}{\omega_n}\int_{|\xi|=1} \Delta_x u(t,x+r\xi)\,dS_\xi = \Delta_x M(u)(t,x,r)\ .\end{aligned}$$

Now use the Darboux equation of Lemma 6.5,

$$\partial_t^2 M(u)(t,x,r) = \big(\partial_r^2 + \frac{n-1}{r}\partial_r\big) M(u)(t,x,r)\ , \tag{6.10}$$

a PDE in the two variables $(t,r) \in \mathbb{R}^2$. This is known as the *Euler–Poisson–Darboux* equation.

Spherical means in $\mathbb{R}^3$. Equation (6.10) is normally posed as an initial value problem

$$\begin{aligned}&\partial_t^2 M(u)(t,x,r) = \big(\partial_r^2 + \frac{n-1}{r}\partial_r\big) M(u)(t,x,r)\ , \\ &M(u)(0,x,r) = M(f)(x,r)\ , \qquad \partial_t M(u)(0,x,r) = M(g)(x,r)\ .\end{aligned} \tag{6.11}$$

The precise methods and the character of the solution depend quite a bit on the spatial dimension under consideration. The most straightforward case is in dimension $n = 3$. In the Euler–Poisson–Darboux equation (6.11), use the substitution $M(u)(t,x,r) \mapsto rM(u)(t,x,r)$, giving

$$\partial_t^2\big(rM(u)(t,x,r)\big) = r\big(\partial_r^2 + \frac{2}{r}\partial_r\big)M(u)(t,x,r) = \partial_r^2\big(rM(u)(t,x,r)\big)\ .$$

Therefore the function $v(t,r) = rM(u)(t,x,r)$ is a solution of the wave equation in one dimension, and the variable x is relegated to the role of a parameter. The solution is given by the d'Alembert formula

$$\begin{aligned}v(t,r) &= rM(u)(t,x,r) \\ &= \frac{1}{2}\big((r+t)M(f)(x,r+t) + (r-t)M(f)(x,r-t)\big) \\ &\quad + \frac{1}{2}\int_{r-t}^{r+t} \rho M(g)(x,\rho)\,d\rho\ .\end{aligned}$$

Since $M(f)(x,r)$ and $M(g)(x,r)$ are even functions under $r \mapsto -r$, then $rM(f)(x,r)$ and $rM(g)(x,r)$ are odd; therefore we may rewrite the above expression, dividing through by r:

$$\begin{aligned}M(u)(t,x,r) &= \frac{1}{2r}\big((t+r)M(f)(x,t+r) - (t-r)M(f)(x,t-r)\big) \\ &\quad + \frac{1}{2r}\int_{t-r}^{t+r} \rho M(g)(x,\rho)\,d\rho\ .\end{aligned}$$

We have used the fact that $\int_{t-r}^{r-t} \rho M(g)(x,\rho)\, d\rho = 0$, which holds because $rM(g)(x,r)$ is odd. Taking the limit as $r \to 0$ and using Proposition 6.6, we recover a representation for the solution $u(t,x)$.

Theorem 6.7 (Kirchhoff's formula). *When $n = 3$, the solution to the wave equation* (6.1) *is given by the expression*

$$u(t,x) = \partial_t\big(tM(f)(x,t)\big) + tM(g)(x,t) \ , \tag{6.12}$$

which is well defined as long as $f(x) \in C^1(\mathbb{R}^3)$ and $g(x) \in C(\mathbb{R}^3)$.

It is quite explicit to see that in terms of pointwise regularity, the solution is generally less smooth than the initial data, for the Kirchhoff formula depends upon the derivative of $f(x)$. Specifically, carrying out the differentiation in (6.12), we obtain

$$u(t,x) = \frac{1}{4\pi t^2} \int_{|x-y|=t} \big(tg(y) + f(y) + t\nabla f(y) \cdot \frac{x-y}{|x-y|}\big)\, dS_y \ .$$

Corollary 6.8. *Given initial data $g(x) \in C^2(\mathbb{R}^3)$ and $f(x) \in C^3(\mathbb{R}^3)$, the spherical means solution given in* (6.12) *is a solution $u(t,x) \in C^2(\mathbb{R}^1_t \times \mathbb{R}^3_x)$.*

The corollary points to the phenomenon of derivative loss in the solution process for the wave equation, where three derivatives of the initial data are required in order to ensure that the solution is classical, namely twice differentiable. This loss of differentiability for $n \geq 2$, due to focusing effects, is the topic of Exercise 6.2. In contrast, when solutions are viewed in the sense of the Sobolev spaces H^s through the energy, there is no loss visible:

$$E(u)(t) = \int \frac{1}{2}(\partial_t u)^2 + \frac{1}{2}|\nabla_x u|^2\, dx = \int \frac{1}{2}g^2 + \frac{1}{2}|\nabla_x f|^2\, dx \ .$$

The Sobolev regularity of the solution $u(t,x)$ is the same as the Sobolev regularity of the initial data.

Spherical means in $\mathbb{R}^n$ for odd n. There are similar expressions for the solution of the wave equation for $x \in \mathbb{R}^n$, for $n \geq 3$ and odd. Returning to the Euler–Poisson–Darboux equation (6.10) for the n-dimensional wave equation

$$\partial_t^2 M(u) = \big(\partial_r^2 + \frac{n-1}{r}\partial_r\big) M(u) \ ,$$

we seek an algebraic reduction to the wave equation in two dimensions in the variables (t,r). This is able to be carried out in the odd-dimensional case.

Proposition 6.9. *Suppose that $k \geq 1$ is an integer and that $h = h(r) \in C^{k+1}(\mathbb{R}^1_+)$. Then*

$$\partial_r^2 \Big(\frac{1}{r}\partial_r\Big)^{k-1}\Big(r^{2k-1}h\Big) = \Big(\frac{1}{r}\partial_r\Big)^{k}\Big(r^{2k}\partial_r h\Big) ,$$

$$\Big(\frac{1}{r}\partial_r\Big)^{k-1}\Big(r^{2k-1}h\Big) = \sum_{j=0}^{k-1} \beta_j^k r^{j+1}\partial_r^j h$$

with the combinatorial coefficients being $\beta_0^k = 1 \cdot 3 \cdot \ldots (2k-1) := (2k-1)!!$.

Proof. An argument by induction will work. □

The result is useful because we may set $n = 2k+1$ and take

$$v(t,x,r) := \Big(\frac{1}{r}\partial_r\Big)^{k-1}\Big(r^{2k-1}M(u)\Big)(t,x,r)$$

for a given solution $u(t,x)$ of (6.1). Then using the identities from Proposition 6.9,

$$\begin{aligned}
\partial_r^2 v &= \partial_r^2 \Big(\frac{1}{r}\partial_r\Big)^{k-1}\Big(r^{2k-1}M(u)\Big) \\
&= \Big(\frac{1}{r}\partial_r\Big)^{k}\Big(r^{2k}\partial_r M(u)\Big) \\
&= \Big(\frac{1}{r}\partial_r\Big)^{k-1}\Big(r^{2k-1}\partial_r^2 M(u) + 2kr^{2k-2}\partial_r M(u)\Big) \\
&= \Big(\frac{1}{r}\partial_r\Big)^{k-1}\Big(r^{2k-1}(\partial_r^2 M(u) + \frac{2k}{r}\partial_r M(u))\Big) \\
&= \Big(\frac{1}{r}\partial_r\Big)^{k-1}\Big(r^{2k-1}\partial_t^2 M(u)\Big) = \partial_t^2 v(t,r) .
\end{aligned}$$

Therefore $v(t,x,r)$ is the quantity that satisfies the one-dimensional wave equation in the variables (t,r), for which we can apply the d'Alembert formula:

$$v(t,x,r) = \frac{1}{2}\big(v(0,r+t) + v(0,r-t)\big) + \frac{1}{2}\int_{r-t}^{r+t} \partial_t v(0,x,\rho)\,d\rho$$

with initial data

$$\begin{aligned}
v(0,x,r) &= \Big(\frac{1}{r}\partial_r\Big)^{k-1}(r^{2k-1}M(f))(x,r) , \\
\partial_t v(0,x,r) &= \Big(\frac{1}{r}\partial_r\Big)^{k-1}(r^{2k-1}M(g))(x,r) .
\end{aligned}$$

One recovers the solution as a limit

$$u(t,x) = \lim_{r\to 0} M(u)(t,x,r) = \lim_{r\to 0} \frac{v(t,x,r)}{\beta_0^k r} \tag{6.13}$$
$$= \frac{1}{(n-2)!!}\left[\partial_t\left(\frac{1}{t}\partial_t\right)^{(n-3)/2}\left(t^{n-2}M(f)(t,x)\right) + \left(\frac{1}{t}\partial_t\right)^{(n-3)/2}\left(t^{n-2}M(g)(t,x)\right)\right] .$$

Theorem 6.10. *For n odd, $f \in C^{(n-1)/2}(\mathbb{R}^n)$, and $g \in C^{(n-3)/2}(\mathbb{R}^n)$, the solution formula* (6.13) *gives a classical solution to the wave equation in $\mathbb{R}^n$.*

It is interesting to quantify the possible loss of smoothness of the solution over the initial data, made clear by formula (6.13). Indeed, the reduction process from $M(u)$ to v involves $k-1 = (n-3)/2$ derivatives, and the limit involves one derivative; therefore in general the solution will be less regular than the initial data by $k = (n-1)/2$ many derivatives.

6.4. Huygens' principle

Huygens' principle is the expression of the principle of finite propagation speed, analogous to the case of the one-dimensional wave equation in Chapter 2. This general form of Huygens' principle is valid for the wave equation in any space dimension and for hyperbolic equations in general. The strong form of Huygens' principle says more than this; it is the property that solutions are supported precisely on the union of light cones which have their vertex on the support of the initial data. There is a difference in dimension concerning this strong form of the Huygens' principle; it holds for odd space-dimensional problems, but not in even dimensions. We state it here for the case of three space dimensions.

Theorem 6.11. *Consider solutions to the wave equation for $x \in \mathbb{R}^3$, and assume that the Cauchy data $f(x)$ and $g(x)$ are compactly supported, such that* $\mathrm{supp}(f) \cup \mathrm{supp}(g) \subseteq B_R(0)$. *Then*

(1) (Huygens' principle). *The solution $u(t,x)$ has its support within the bounded region $B_{R+|t|}(0)$:*

$$\mathrm{supp}(u(t,\cdot)) \subseteq B_{R+|t|}(0) .$$

(2) (Strong Huygens' principle). *Additionally, for $|t| > R$, for any space-time point (t,x) in the region inside the light cones given by $\{(t,x) : |x| \le |t| - R\}$, again the solution vanishes; $u(t,x) = 0$.*

Proof. (1) *of Huygens' principle.* The result follows from the form of the Kirchhoff formula. Express the solution $u(t,x)$ in terms of spherical means over spheres of radius $|t|$ in the initial hyperplane $\{t = 0\}$ that are of the form $\{y : |x - y| = |t|\}$ (the intersection of the backwards light cone emanating from (t, x) and the Cauchy hypersurface). If $|x| > R+|t|$, then the associated sphere does not intersect $B_R(0)$, which contains the support of the initial data, and hence the solution at that space-time point vanishes.

(2) *of the strong Huygens' principle.* When $|t| > R$ and $|x| < |t| - R$, again the backwards light cone emanating from the point (t, x) does not intersect the support of the initial data, in this case because the sphere $\{y : |x - y| = |t|\}$ is too big and has passed outside of $B_R(0)$.

One notes that this proof is geometric, only depending upon the fact that the solution is represented as a spherical mean. Therefore the result holds for solutions of the wave equation in arbitrary odd dimensions, as exhibited in (6.13). □

In fact a more precise statement is true; at times $t > 0$, the solution $u(t, x)$ is supported within the region consisting of the union of light cones

$$LC_+(t, x) = \{(t, y) : t^2 - |y - x|^2 = 0, \, t > 0\}$$

whose vertices x lie in the set $\text{supp}(f) \cup \text{supp}(g)$. A space-time picture of the support of the solution is depicted in Figure 2.

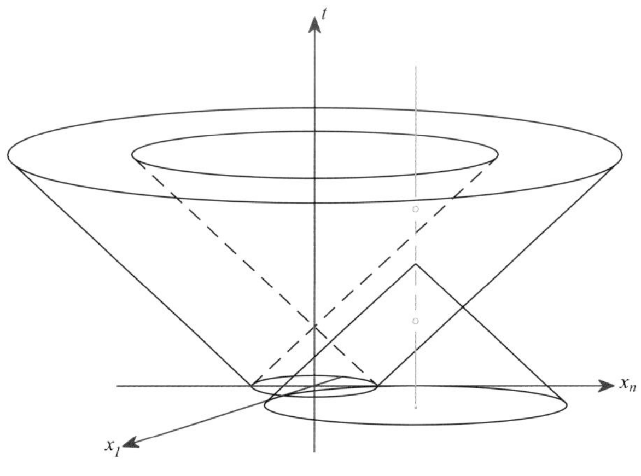

Figure 2. Space-time picture of the support of a solution. The vertical green line is the world line of a stationary observer.

The finite propagation speed property is sometimes known as the weak Huygens' principle; it is a central feature of hyperbolic equations. It implies that in particular a signal will travel with finite velocity. That is, for an observer standing still at point $x_0 \in \mathbb{R}^n$, $|x_0| > R$, the solution satisfies $u(x_0, t) = 0$ until $|t| + R > |x_0|$.

Corollary 6.12. *If the initial data $f(x), g(x)$ vanish on $B_R(0)$, then for $0 < |t| < R$, the solution $u(t,x)$ must vanish on the cone $\{(t,x) : |x| < R - |t|\}$. In particular if f, g are identically zero, then $u(t,x)$ will also be zero.*

Proof. This is a local uniqueness theorem. Suppose that $\operatorname{supp}(f) \cup \operatorname{supp}(g) \subseteq H_{a,v} = \{x \in \mathbb{R}^n : a \leq v \cdot x\}$, a half space; then $\operatorname{supp}(u(t,x)) \subseteq H_{a+|t|,v}$. Therefore no data in any of the enveloping half space $H_{R-|t|,v}$, $|v| = 1$ to $B_{R-|t|}(0)$ can propagate into the cone pictured in Figure 3. □

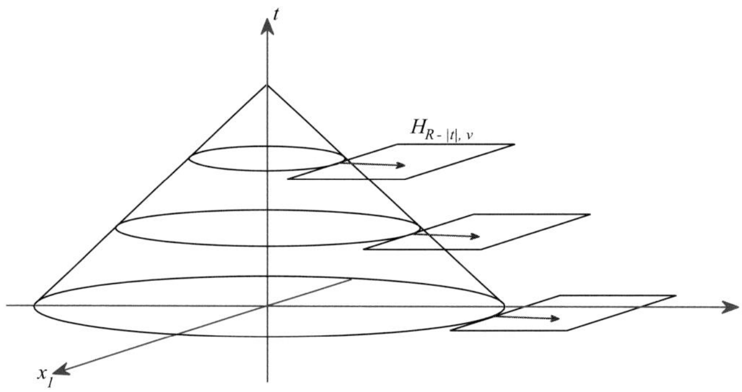

Figure 3. The cone of uniqueness.

Definition 6.13. (i) The *domain of influence* of a set $A \subseteq \mathbb{R}^n$ is the region in $\mathbb{R}^{n+1}$ in which a solution $u(x,t)$ of the wave equation can be affected by data in A.

(ii) The *domain of dependence* of a set $B \subseteq \mathbb{R}^{n+1}$ is the region $A \subseteq \mathbb{R}^n$ in which data can influence the solution in B.

The results of Theorem 6.17 and Corollary 6.12 imply the following pictures for the domains of influence and dependence of solutions of the wave equation, given respectively in Figures 4 and 5.

6.5. Paley–Wiener theory

It is evident that the complex exponentials $e^{-i\xi \cdot x}$, with $\xi \in \mathbb{R}^n$ and $x \in \mathbb{R}^n$, extend as holomorphic functions $e^{-i\zeta \cdot x}$ to $\zeta \in \mathbb{C}^n$. More specifically to its character, the extension is defined over all of $\mathbb{C}^n$, meaning that it is *entire*, and furthermore the extension has bounds on its growth at infinity of the form

$$|e^{-i\zeta \cdot x}| \leq e^{|x||im(\zeta)|}.$$

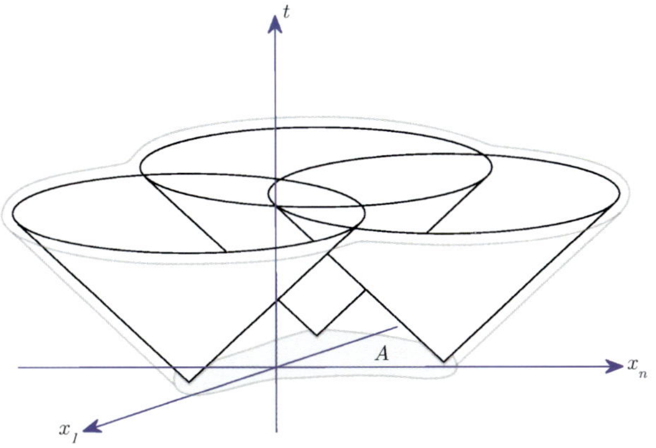

Figure 4. Domain of influence of the set $A \subseteq \mathbb{R}^n$.

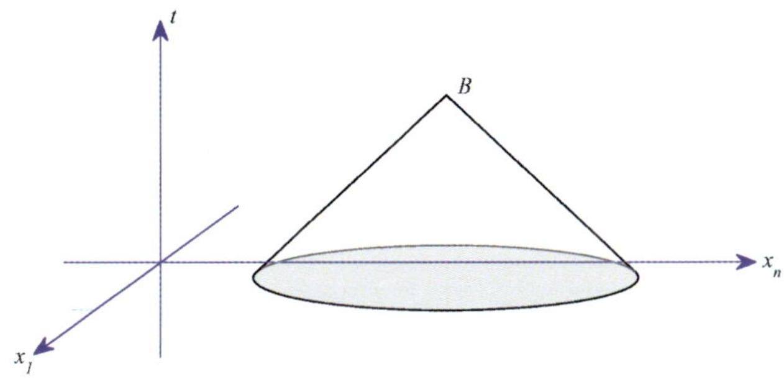

Figure 5. Domain of dependence of the point B.

Definition 6.14. An entire function $g(\zeta)$, $\zeta \in \mathbb{C}^n$, is of *exponential type* R if it satisfies the estimates

$$(1 + |\zeta|^2)^{N/2} |g(\zeta)| \le C_N e^{R|im(\zeta)|} \tag{6.14}$$

for all $N \in \mathbb{N}$.

Examples of the behavior are given by the Fourier transform of C_0^∞ functions, as we will describe presently. Indeed, the factor $(1 + |\zeta|^2)^{N/2}$ is present to ensure that all such functions have smooth, well-behaved Fourier transforms. The complex exponential above is not strictly speaking of exponential type because it only satisfies (6.14) for the case $N = 0$. Another example of an entire function, which, however, is not of exponential type, is

$$g(\zeta) = e^{-\zeta^2/2} = e^{-\frac{1}{2}(\xi^2 - \eta^2) - i\xi\eta}$$

for $\zeta = \xi + i\eta \in \mathbb{C}$. Given any function $f \in \mathcal{S}$, its Fourier transform can be viewed as a superposition of complex exponentials

$$\hat{f}(\xi) = \frac{1}{\sqrt{2\pi}^n} \int_{\mathbb{R}^n} e^{-i\xi \cdot x} f(x)\, dx \ ,$$

where each complex exponential extends to an entire function on $\mathbb{C}^n$ but the function $\hat{f}(\xi)$ itself need not necessarily have a holomorphic extension off of the real subspace $\mathbb{R}^n \subseteq \mathbb{C}^n$ at all. However, if a function $g(\xi)$ is such that $g(\xi) = \hat{f}(\xi)$ with $f(x) \in C_0^\infty(\mathbb{R}^n)$, in particular if $f(x)$ has compact support, then $g(\xi)$ will have an entire holomorphic extension.

Proposition 6.15. *Suppose that $f \in C_0^\infty(\mathbb{R}^n)$, with $\mathrm{supp}(f) \subseteq \{x \in \mathbb{R}^n : |x| \leq R\}$. Then $\hat{f}(\xi) = g(\xi)$ extends to an entire function, which is of exponential type R.*

Proof. For each $\zeta \in \mathbb{C}^n$, the integral

$$g(\zeta) = \frac{1}{\sqrt{2\pi}^n} \int_{B_R(0)} e^{-i\zeta\cdot x} f(x) dx$$

converges absolutely, uniformly over bounded sets of ζ. To check that $g(\zeta)$ is holomorphic, we simply test the Cauchy–Riemann equations:

$$\partial_{\bar{\zeta}} g(\zeta) = \frac{1}{\sqrt{2}}(\partial_\xi + i\partial_\eta) g(\zeta) = \frac{1}{\sqrt{2\pi}^n} \int \partial_{\bar{\zeta}}(e^{-i\zeta\cdot x}) f(x)\, dx = 0 \ .$$

Lastly, for α a multi-index with $|\alpha| = N$, we need to estimate the integrals

$$\begin{aligned} \left| \zeta^\alpha \int_{\mathbb{R}^n} e^{-i\zeta\cdot x} f(x) dx \right| &= \left| \int_{B_R(0)} e^{-i\zeta\cdot x} \left(\frac{1}{i} \partial_x \right)^\alpha f(x) dx \right| \\ &\leq \int_{B_R(0)} e^{|x||im(\zeta)|} |\partial_x^\alpha f(x)| dx \leq C_{|\alpha|} e^{R|im(\zeta)|}. \end{aligned}$$

□

The content of Paley–Wiener theory is to say that the converse also holds.

Theorem 6.16 (Paley–Wiener)**.** *Suppose that $g(\zeta)$ is an entire function of exponential type R. Then there is a function $f(x) \in C_0^\infty(\mathbb{R}^n)$ with $\mathrm{supp}(f) \subseteq B_R(0)$ such that*

$$g(\zeta) = \hat{f}(\zeta) = \frac{1}{\sqrt{2\pi}^n} \int e^{-i\zeta\cdot x} f(x)\, dx \ .$$

Proof. It is natural to restrict g to $\xi \in \mathbb{R}^n$, and define

$$f(x) = \frac{1}{\sqrt{2\pi}^n} \int_{\mathbb{R}^n} e^{i\xi\cdot x} g(\xi) d\xi.$$

Since $|g(\xi)| \leq C_N (1 + |\xi|^2)^{-N/2}$ as in (6.14), this integral is absolutely convergent. We may also take an arbitrary number of derivatives of $f(x)$,

$$\partial_x^\beta f(x) = \frac{1}{\sqrt{2\pi}^n} \int_{\mathbb{R}^n} e^{i\xi\cdot x} (-i\xi)^\beta g(\xi)\, d\xi \ ,$$

and the integral remains absolutely convergent. Thus $f \in C^\infty(\mathbb{R}^n)$ and the question that remains has to do with its support.

Using the fact that $g(\zeta)$ is holomorphic, consider deformations of the region of integration off of the real axis in the complex space $\mathbb{C}^n$,

$$\frac{1}{\sqrt{2\pi}^n} \int e^{i((\xi_1+i\eta_1)x_1+\xi'\cdot x')} g(\xi_1+i\eta_1, \xi')\, d\xi_1 d\xi'$$

for $\eta_1 \in \mathbb{R}$. This is independent of η_1, as can be shown by Cauchy's theorem, taking the limit of an integral in $\zeta_1 = \xi_1 + i\eta_1$ over the contour given by Figure 6 and letting $T \to +\infty$. The decay condition (6.14) implies that there

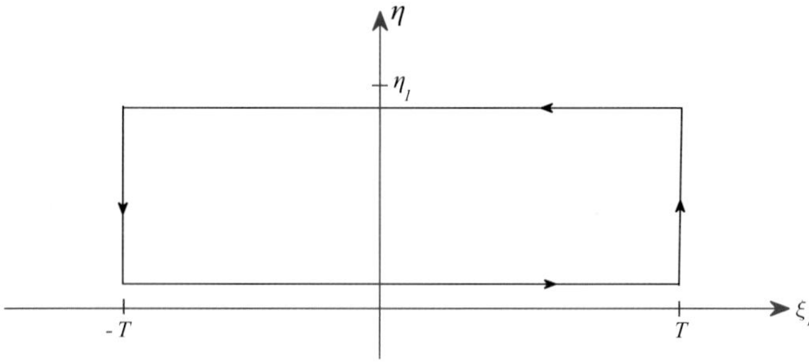

Figure 6. The contour.

are no contributions from the boundaries $\xi_1 = \pm T$ in the limit. Repeating this argument in all variables, we show that for any desired $\eta \in \mathbb{R}^n$,

$$f(x) = \frac{1}{\sqrt{2\pi}^n} \int_{\mathbb{R}^n} e^{i(\xi+i\eta)\cdot x} g(\xi+i\eta)\, d\xi \ .$$

Now fix $x \neq 0$ in $\mathbb{R}^n$, and choose the particular $\eta = \lambda \frac{x}{|x|}$ with a real parameter $\lambda > 0$. Then

$$\begin{aligned} |f(x)| &\le \frac{1}{\sqrt{2\pi}^n} \left| \int_{\mathbb{R}^n} e^{i\xi\cdot x - \lambda|x|} g\left(\xi + i\lambda\frac{x}{|x|}\right) d\xi \right| \\ &\le \frac{1}{\sqrt{2\pi}^n} \int_{\mathbb{R}^n} e^{-\lambda|x|} \left| g\left(\xi + i\lambda\frac{x}{|x|}\right)\right| d\xi \\ &\le C_N \int_{\mathbb{R}^n} (1+|\xi|^2)^{-N/2} e^{-\lambda|x|+\lambda R} d\xi \ . \end{aligned}$$

Now suppose that $|x| > R$; as $\lambda \to +\infty$, the RHS tends to zero. Thus we have shown that $f(x) = 0$. This proves that indeed $\mathrm{supp}(f) \subseteq B_R(0)$. □

Concerning the class of distributions mentioned in Chapter 5, there is a similar theory for distributions of compact support, and their Fourier transforms are entire functions of exponential type. Conversely if $g(\zeta)$ is an entire function which for some N satisfies the estimate

$$|g(\zeta)| \le C(1+|\zeta|^2)^{N/2} e^{R|im(\zeta)|} \ , \tag{6.15}$$

then there is a distribution $f \in \mathcal{D}'$ with $\mathrm{supp}(f) \subseteq B_R(0)$ such that the (generalized) Fourier transform of f is g.

Huygens' principle revisited. Returning to the discussion of solutions of the wave equation, our Fourier integral expression is that

$$u(x,t) = \frac{1}{\sqrt{2\pi}^n} \int e^{i\xi\cdot x} \left(\cos(|\xi|t)\hat{f}(\xi) + \frac{\sin(|\xi|t)}{|\xi|}\hat{g}(\xi) \right) d\xi \ .$$

Let us suppose that the initial data is of compact support:

$$\mathrm{supp}(f) \cup \mathrm{supp}(g) \subseteq B_R(0) = \{x \in \mathbb{R}^n : |x| < R\} \ .$$

Then by Proposition 6.15, both $\hat{f}(\xi)$ and $\hat{g}(\xi)$ extend to entire functions on $\mathbb{C}^n$ of exponential type R. Let us examine the Fourier transform of the solution

$$\hat{u}(\xi,t) = \cos(|\xi|t)\hat{f}(\xi) + \frac{\sin(|\xi|t)}{|\xi|}\hat{g}(\xi) \ .$$

The individual functions $\cos(\sqrt{\zeta^2}t)\hat{f}(\zeta)$ and $\frac{\sin(\sqrt{\zeta^2}t)}{\sqrt{\zeta^2}}\hat{g}(\zeta)$ are entire functions of $\zeta \in \mathbb{C}^n$, where we are using the notation that $\zeta^2 = \zeta_1^2 + \zeta_2^2 + \cdots + \zeta_n^2$ for $\zeta_j \in \mathbb{C}$. Furthermore,

$$\left|\cos(\sqrt{\zeta^2}t)\right| \leq e^{|im(\zeta)||t|}, \qquad \left|\frac{\sin(\sqrt{\zeta^2}t)}{\sqrt{\zeta^2}}\right| \leq e^{|im(\zeta)||t|} \ ;$$

therefore the two products $\cos(\sqrt{\zeta^2}t)\hat{f}(\zeta)$ and $\left(\sin(\sqrt{\zeta^2}t)/\sqrt{\zeta^2}\right)\hat{g}(\zeta)$ are of exponential type $(R + |t|)$. We can conclude that the solution $u(x,t)$ of the wave equation is the Fourier transform of an entire function of exponential type $(R + |t|)$, and thus by the Paley–Wiener theorem, the solution $u(x,t)$ has its support in the ball $B_{R+|t|}(0)$ of radius $R + |t|$. This is a second proof of Huygens' principle for the wave equation.

Theorem 6.17 (Huygens' principle again). *A solution of the wave equation* (6.1) *with initial data with support such that* $\mathrm{supp}(f) \cup \mathrm{supp}(g) \subseteq B_R(0)$ *satisfies at nonzero times* $t \in \mathbb{R}$

$$\mathrm{supp}(u(t,x)) \subseteq B_{R+|t|}(0) \ .$$

This is a proof that is valid for all spatial dimensions $n \geq 1$, independent of the parity of the dimension. Because of this independence, techniques based on the Fourier transform are, however, not so well adapted for a proof of the strong Huygens' principle that holds in odd space dimensions, as stated in Theorem 6.11(2) for $n = 3$, but not in even dimensions.

6.6. Lagrangians and Hamiltonian PDEs

Many PDEs that arise in physics are derived from a Lagrangian functional, following a formal but nevertheless compelling analogy of the *principle of least action* of classical mechanics. The wave equation is a typical example. The *Lagrangian* function for the wave equation (6.1) is defined by

$$L := \int_{\mathbb{R}^n} \frac{1}{2}(\partial_t u)^2 - \frac{1}{2}|\nabla u|^2 \, dx \ . \tag{6.16}$$

This Lagrangian has an associated *action integral* given by

$$S := \int_0^T L(u, \partial_t u)\, dt = \int_0^T \int_{\mathbb{R}^n} \frac{1}{2}(\partial_t u)^2 - \frac{1}{2}|\nabla u|^2 \, dxdt \ . \tag{6.17}$$

From this action integral S, the wave equation arises from the principle of least action, which dictates that the motion of a Lagrangian system is a stationary point of the action integral. A stationary point of the action over the time interval $0 \leq t \leq T$ is a function $u(t,x)$ which satisfies $\delta S = 0$ for all admissible variations $v(t,x) = \delta u(t,x)$, where $\delta u(t,x)$ denotes a small but arbitrary variation of the function $u(t,x)$. Admissible variations are smooth and are such that $v(0,x) = v(T,x) = 0$, so that $u(t,x)$ and $u(t,x) + v(t,x)$ have the same initial and final states over the time interval $[0,T]$. In the case of the Lagrangian L for the wave equation, a stationary point of the action satisfies

$$\begin{aligned}
\delta S &= \int_0^T \int_{\mathbb{R}^n} \partial_t u \partial_t v - \nabla_x u \cdot \nabla_x v \, dxdt \\
&= -\int_0^T \int_{\mathbb{R}^n} (\partial_t^2 u - \Delta u) v \, dxdt + \int_{\mathbb{R}^n} \partial_t u v \, dx \Big|_{t=0}^{T} \ .
\end{aligned}$$

From the assumption that $v(0,x) = v(T,x) = 0$, the last term of the RHS vanishes. The conclusion is that because $v(t,x)$ is otherwise arbitrary, we must have that

$$\partial_t^2 u - \Delta u = 0 \ .$$

These are the *Euler–Lagrange* equations for the action (6.17). Notice that, ironically, the formal principle of stationary action allows us to give an initial *position* $u(0,x) = f(x)$ but does not allow for setting the initial *momentum* $\partial_t u(0,x) = g(x)$. Furthermore this formal principle asks us to specify the final position $u(T,x)$, so it is not in fact compatible with an initial value problem. Nonetheless, the principle of least action, or more generally of stationary action, remains a guiding principle for many equations in physics, while on a rigorous mathematical level the principle remains a formal one in this and other cases.

A Lagrangian L and subsequently an action integral S can be defined for more general systems; indeed, this is how wave equations are derived in most problems in physics. Consider the more general Lagrangian

$$L = \int_{\mathbb{R}^n} \frac{1}{2}(\partial_t u)^2 - G(u, \nabla_x u)\, dx \tag{6.18}$$

whose associated action is given by

$$S = \int_0^T L\, dt = \int_0^T \int_{\mathbb{R}^n} \frac{1}{2}(\partial_t u)^2 - G(u, \nabla_x u)\, dxdt\ .$$

Suppose that the field $u(t,x)$ is a stationary point of the action:

$$\begin{aligned} 0 = \delta S &= \frac{d}{d\varepsilon}\Big|_{\varepsilon=0} \int_0^T \int_{\mathbb{R}^n} \frac{1}{2}(\partial_t u + \varepsilon \partial_t v)^2 - G(u + \varepsilon v, \nabla_x(u + \varepsilon v))\, dxdt \\ &= \int_0^T \int_{\mathbb{R}^n} \partial_t u \partial_t v - \partial_{\nabla u} G(u, \nabla_x u) \cdot \nabla_x v - \partial_u G(u, \nabla_x u) v\, dxdt \\ &= \int_0^T \int_{\mathbb{R}^n} \big[-\partial_t^2 u + \nabla_x \cdot \partial_{\nabla u} G(u, \nabla_x u) - \partial_u G(u, \nabla_x u)\big] v\, dxdt \\ &\quad + \int_{\mathbb{R}^n} \partial_t u v\, dx\Big|_{t=0}^T\ . \end{aligned}$$

The notation is that the nonlinear function $G = G(u, V)$ depends upon the variables u as well as the n components of $V = \nabla u$. The notation for its partial derivatives is that $\partial_{\nabla u} G(u, \nabla_x u) = \partial_V G(u, V)|_{V=\nabla_x u}$. The final term vanishes because $v(t,x)$ is an admissible variation. Since v is otherwise arbitrary, the field $u(t,x)$ must satisfy the Euler–Lagrange equations

$$\partial_t^2 u - \nabla_x \cdot \partial_{\nabla u} G(u, \nabla_x u) + \partial_u G(u, \nabla_x u) = 0\ .$$

This is a hyperbolic equation if the matrix of partial derivatives of G with respect to the variables V is positive definite. In the example of the wave equation, $G(u, \nabla u) = \frac{1}{2}|\nabla u|^2$, and $\partial_{\nabla u}^2 G = I$.

In this formal treatment of the derivation of field theories, there is another useful analogy with classical mechanics. Given a Lagrangian L, there is a transformation of the Euler–Lagrange equations to a Hamiltonian system, in our case to a system of Hamiltonian PDEs. This is illustrated in the example Lagrangian

$$L(u, \dot{u}) = \int_{\mathbb{R}^n} \frac{1}{2}\dot{u}^2 - G(\nabla_x u)\, dx\ , \tag{6.19}$$

where for clarity we have simplified the Lagrangian (6.18) above, and we are using the notation that $\partial_t u = \dot{u}$. The action integral is as before

$$S = \int_0^T L\, dt,$$

and as above the principle of stationary action gives the Euler–Lagrange equations

$$-\partial_t \delta_{\dot{u}} L + \delta_u L = 0 \ , \tag{6.20}$$

where $\delta_u L = \nabla_x \cdot \partial_{\nabla u} G(\nabla_x u)$ and $\delta_{\dot{u}} L = \dot{u}$. In general a Lagrangian may depend explicitly on time $L(\dot{u}, u, t)$, but in many cases, such as the one at hand, it describes a physical process whose properties do not change with time, and L is independent of t. In this situation, the Euler–Lagrange equations exhibit a conservation law. This can be seen from the following computation:

$$\begin{aligned}\frac{d}{dt} L &= \int \big(\delta_{\dot{u}} L \, \ddot{u} + \delta_u L \, \dot{u}\big) \, dx = \int \big(\delta_{\dot{u}} L \, \ddot{u} + \partial_t(\delta_{\dot{u}} L) \, \dot{u}\big) \, dx \\ &= \frac{d}{dt} \int \delta_{\dot{u}} L \, \dot{u} \, dx \ .\end{aligned}$$

We used the Euler–Lagrange equations (6.20) in the last equality of the first line. Therefore the conservation law is evident, namely

$$\frac{d}{dt}\Big(\int \delta_{\dot{u}} L \, \dot{u} \, dx - L\Big) = 0 \ . \tag{6.21}$$

In example (6.19), this conservation law is

$$\frac{d}{dt} \int_{\mathbb{R}^n} \frac{1}{2} (\partial_t u)^2 + G(\nabla_x u) \, dx = 0 \ .$$

Definition 6.18. (i) Define the *Hamiltonian* of system (6.20) by

$$H := \int \delta_{\dot{u}} L \, \dot{u} \, dx - L \ .$$

(ii) The *conjugate momentum* of the field u is defined to be

$$p := \delta_{\dot{u}} L \ ,$$

giving $p = p(\dot{u}, u)$.
(iii) If the relationship between $\dot{u}$ and p can be inverted to obtain $\dot{u} = \dot{u}(p, u)$, at least locally, then the mapping $\dot{u} \to p$ is called the *Legendre transform*. Using this mapping, we may rewrite the Hamiltonian

$$H = \int \dot{u} \, p \, dx - L = H(u, p)$$

in terms of the new Hamiltonian variables (u, p).

The Legendre transform offers an elegant way to transform a second-order equation (in time) to a first-order system of equations.

Theorem 6.19. *The Euler–Lagrange equations (6.20) for $u = u(t,x)$ are equivalent in the new variables $(u(t,x), p(t,x))$ to the system*

$$\partial_t u = \delta_p H \ , \qquad \partial_t p = -\delta_u H \ . \tag{6.22}$$

The system of equations (6.22) is known as *Hamilton's canonical equations* for the evolution equations described by H.

Proof. The formal equivalence of (6.20) and (6.22) is a general fact. Firstly,

$$H = \int \dot{u}\, p\, dx - L$$

so that $\dot{u} = \delta_p H$. Secondly, we note that $\delta_u H = -\delta_u L$ so that

$$\dot{p} = \frac{d}{dt}\big(\delta_{\dot{u}} L\big) = \delta_u L = -\delta_u H \ .$$

□

Exhibiting this transformation in the setting of the wave equation, we have

$$L = \int_{\mathbb{R}^n} \frac{1}{2}\dot{u}^2 - \frac{1}{2}|\nabla_x u|^2 \, dx \ ,$$

from which the Legendre transform gives $p = \delta_{\dot{u}} L = \dot{u}$. Then

$$H = \int \dot{u}\, p\, dx - L(u, \dot{u}) = \int \dot{u}^2 \, dx - L = \int \frac{1}{2}p^2 + \frac{1}{2}|\nabla_x u|^2 \, dx \ .$$

Hamilton's canonical equations are then

$$\partial_t u = \delta_p H = p \ , \qquad \partial_t p = -\delta_u H = \Delta u \ , \tag{6.23}$$

which is of course the wave equation presented as a first-order system of equations. The energy functional $E(u)$ for the wave equation is the Hamiltonian $H(u,p)$ for the system.

Hamiltonian PDEs. The above formal exposition is a lead-in to study other PDEs that can be posed in the form of Hamiltonian systems. Consider systems of equations of the form

$$\partial_t z = J \mathrm{grad}_z H \ , \tag{6.24}$$

where z is a vector valued function in a Hilbert space $\mathcal{H}$, and the matrix $J = -J^T$ is skew symmetric on $\mathcal{H}$. The gradient $\mathrm{grad}_z H$ of a functional is defined using the inner product of $\mathcal{H}$ in this way:

$$\delta_z H \cdot v = \frac{d}{ds}\Big|_{s=0} H(z + sv) := \langle \mathrm{grad}_z H, v \rangle_{\mathcal{H}} \ .$$

Rewriting the wave equation (6.23) as

$$\partial_t \begin{pmatrix} u \\ p \end{pmatrix} = \begin{pmatrix} p \\ \Delta u \end{pmatrix} = \begin{pmatrix} 0 & I \\ -I & 0 \end{pmatrix} \begin{pmatrix} -\Delta u \\ p \end{pmatrix} := J\mathrm{grad}_{(u,p)} H \ , \tag{6.25}$$

the wave equation takes this form, with $\mathcal{H} = L^2_u \oplus L^2_p$ and

$$J := \begin{pmatrix} 0 & I \\ -I & 0 \end{pmatrix} .$$

We note that this matrix J satisfies $J = -J^T$ on $\mathcal{H}$.

Proposition 6.20 (Conservation of energy). *The Hamiltonian $H(u,p)$ is a conserved quantity for solutions of* (6.25).

Proof. This can reasonably be called the law of conservation of energy as the Hamiltonian function is often, not always, the energy of the system being considered. The following classical calculation using the chain rule makes the assumption that solutions to (6.24) exist. Indeed, from (6.24),

$$\frac{d}{dt} H(z(t)) = \langle \mathrm{grad}_z H, \dot{z} \rangle_{\mathcal{H}} = \langle \mathrm{grad}_z H, J\mathrm{grad}_z H \rangle_{\mathcal{H}} = 0 \ ,$$

where we have used the skew symmetry of the matrix J. □

Pursuing our analogy with ODEs, this is a treatment of initial value problems for PDEs that are evolution equations as dynamical systems in a function space, naturally a space of infinite dimensions. A number of other PDEs of interest can be viewed as Hamiltonian systems in infinitely many variables. Several examples follow.

1. Nonlinear wave equations. Semilinear hyperbolic equations of this form can be treated as Hamiltonian PDEs. Define the Hamiltonian to be

$$H(u,p) = \int_{\mathbb{R}^n} \frac{1}{2} p^2 + \frac{1}{2} |\nabla u|^2 + G(x,u) \, dx \ .$$

Then the gradient of H is given by

$$\mathrm{grad} H(u,p) = \begin{pmatrix} -\Delta u + \partial_u G(x,u) \\ p \end{pmatrix} ,$$

and therefore the equations of motion are posed as

$$\partial_t \begin{pmatrix} u \\ p \end{pmatrix} = J \mathrm{grad} H \ ,$$

which is a first-order system equivalent to the nonlinear wave equation

$$\partial_t^2 u = \Delta u - \partial_u G(x,u) \ .$$

2. Nonlinear Schrödinger equations. The Hamiltonian for this set of equations is in the form

$$H(\psi,\bar\psi) = \int \frac{1}{2}|\nabla\psi|^2 + Q(x,\psi,\bar\psi)\,dx \ ,$$

where $Q(x,\psi,\xi) : \mathbb{R}_x \times \mathbb{C}_\psi \times \mathbb{C}_\xi \to \mathbb{C}$ has the property that $Q(x,\psi,\bar\psi)$ is real valued. Then $\delta_{\bar\psi} H(\psi,\bar\psi) = -\frac{1}{2}\Delta_x\psi + \partial_{\bar\psi}Q$, and setting $J = iI$ (which is a nondegenerate, skew symmetric operator),

$$\begin{aligned}\partial_t\psi &= J\delta_{\bar\psi}H \\ &= i(-\frac{1}{2}\Delta\psi + \partial_{\bar\psi}Q(x,\psi,\bar\psi)) \ .\end{aligned}$$

When $Q = \pm|\psi|^4$, this equation is the well-known cubic nonlinear Schrödinger equation, where the $+$ sign is the defocusing case and the $-$ sign is the focusing case.

3. Korteweg–de Vries equation (KdV). This famous dispersive equation first arose in the nineteenth century as a model of waves in the free surface of water in a canal. Currently it is used in modeling numerous phenomena including tsunami propagation. It is also well known as a PDE that is a completely integrable Hamiltonian system, where the study of the phase space of solutions has led to discoveries as far ranging as algebraic geometry and inverse spectral theory. The Hamiltonian is

$$H(q) = \int_{-\infty}^{\infty} \frac{1}{12}(\partial_x q)^2 + G(q)\,dx \ .$$

As above, the gradient of $H(q)$ is given by

$$\delta_q H = -\frac{1}{6}\partial_x^2 q + \partial_q G(q) \ .$$

Setting $J = \partial_x$, which again is a skew symmetric operator, we arrive at Hamilton's canonical equations in the form

$$\begin{aligned}\partial_t q &= \partial_x(-\frac{1}{6}\partial_x^2 q + \partial_q G(q)) \\ &= -\frac{1}{6}\partial_x^3 q + G''(q)\partial_x q \ .\end{aligned}$$

The most well-known versions of the KdV equation are when $G(q) = \frac{1}{3}q^3$ and when $G(q) = \frac{1}{4}q^4$.

Exercises: Chapter 6

Exercise 6.1. This problem concerns the Klein–Gordon equation

$$\partial_t^2 u - \Delta u + m^2 u = 0 \ ,$$

where in physics applications the parameter m usually represents mass. This equation, like the wave equation, is usually posed as an initial value problem with initial data

$$u(0,x) = f(x) \ , \quad \partial_t u(0,x) = g(x) \ .$$

Show that the following energy function is a conserved quantity:

$$E_{KG}(u) = \frac{1}{2}\int_{\mathbb{R}^n} (\partial_t u)^2 + |\partial_x u|^2 + m^2 u^2 \, dx \ .$$

Can you find other conserved quantities for this equation?

Exercise 6.2 (Focusing singularity of solutions of the wave equation in $\mathbb{R}^3$ (d'après John [**9**])). (i) Suppose that the initial data for the wave equation in three dimensions has spherically symmetric data:

$$f(x) = f(r) \ , \quad g(x) = g(r) \ , \qquad r^2 = x_1^2 + x_2^2 + x_3^2 \ .$$

Show that the general solution can be expressed as

$$u(t,r) = \frac{1}{r}\big(F(r+t) + G(r-t)\big) \ ;$$

that is, it consists of an incoming wave and an outgoing wave.

(ii) With the special initial data $u(0,r) = 0$, $\partial_t u(r) = g(r)$ with $g(r)$ an even function of r, then

$$u(t,r) = \frac{1}{2r}\int_{r-t}^{r+t} \rho g(\rho)\, d\rho \ .$$

(iii) Set the initial data to be

$$g(r) = 1 \ , \quad 0 \le r < 1 \ , \quad g(r) = 0 \ , \quad 1 \le r \ .$$

Show that $u(t,r)$ is continuous for $|t| < 1$ but at time $t = 1$ exhibits a jump discontinuity. This is due to the focusing of the singularity in $\partial_t u(0,r)$ given at $t = 0$.

Exercise 6.3. This problem addresses the decay rate of solutions of the wave equation in $\mathbb{R}^3$. Suppose that the initial data $(f(x), g(x)) \in C^1(\mathbb{R}^3) \times C(\mathbb{R}^3)$ and that it is supported in the bounded set $B_1(0)$. By inspecting the Kirchhoff formula for the solution $u(t,x)$, show that

$$|u(t,x)| \le \frac{C}{|t|}$$

for some constant C, which can be quantified using $\|f\|_{C^1(\mathbb{R}^3)}$ and $\|g\|_{C(\mathbb{R}^3)}$.

Exercise 6.4 (Global existence with small initial data for certain nonlinear wave equations). This question is to show that certain nonlinear wave equations possess smooth solutions for all $t \in \mathbb{R}$; see [**12**]. This contrasts with other cases where solutions form singularities in finite time. Consider the equation

$$\begin{aligned} \partial_t^2 v - \Delta v + (\partial_t v)^2 - |\nabla v|^2 = 0 \ , \\ v(0,x) = f(x) \ , \quad \partial_t v(0,x) = g(x) \ , \end{aligned} \tag{6.26}$$

with $(f(x), g(x)) \in C^1(\mathbb{R}^3) \times C(\mathbb{R}^3)$ supported in a compact set.

(i) Setting $u = e^v - 1$, show that $u(t,x)$ satisfies the wave equation

$$\begin{aligned} &\partial_t^2 u - \Delta u = 0 \ , \\ &u(0,x) = e^{f(x)} - 1 := F(x) \ , \quad \partial_t u(0,x) = g(x) e^{f(x)} := G(x) \ , \\ &\quad (F(x), G(x)) \in C^1(\mathbb{R}^3) \times C(\mathbb{R}^3) \ . \end{aligned}$$

Explain why (F, G) has compact support.

(ii) Restricting our attention to the case $x \in \mathbb{R}^3$, show that for sufficiently small $\|F\|_{C^1(\mathbb{R}^3)}$, $\|G\|_{C(\mathbb{R}^3)}$, the solution $u(t,x)$ is bounded by

$$|u(t,x)| < 1 \ ,$$

using the result of Exercise 6.3.

In this case the transformation $v \mapsto u$ is invertible for all $(t,x) \in \mathbb{R}^1_t \times \mathbb{R}^3_x$, giving rise to a global solution $v(t,x)$ of equation (6.26).

Exercise 6.5 (Method of descent for the wave equation for $x \in \mathbb{R}^2$). (i) Show that if $x \in \mathbb{R}^3$ but the Cauchy data for the wave equation only depends upon (x_1, x_2), namely

$$f = f(x_1, x_2) \ , \qquad g = g(x_1, x_2) \ , \tag{6.27}$$

then the solution of the wave equation in $\mathbb{R}^1_t \times \mathbb{R}^3_x$ is also independent of x_3,

$$u = u(t, x_1, x_2) \ ,$$

and therefore $u(t, x_1, x_2)$ satisfies the wave equation in two space dimensions:

$$\partial_t^2 u - (\partial_{x_1}^2 + \partial_{x_2}^2) u = 0 \ .$$

(ii) Use the Kirchhoff formula to express the solution to the wave equation in $\mathbb{R}^3$ for data satisfying (6.27).

(iii) In the expression in (ii), reparametrize the spherical integrals by their projection onto the (x_1, x_2) plane, e.g.,

$$\begin{aligned} &\int_{\mathbb{S}^2 : |x-y|=t} f(y) \, dS_y \\ &= \iint_{|(x_1-y_1, x_2-y_2)|<t} f(y_1, y_2) \sqrt{t^2 - ((x_1-y_1)^2 + (x_2-y_2)^2)} \, dy_1 dy_2 \ , \end{aligned}$$

which gives a general formula in $\mathbb{R}^2$ for the solution of the wave equation

$$\Box u = 0 \ .$$

(iv) Describe the nature of this solution in the case that the support of f and g as functions on $\mathbb{R}^2$ is compact, say, supported in the ball $B_R(0) = \{|(x_1, x_2)| < R\}$. In particular comment on Huygens' principle. Does the solution violate the strong form of Huygens' principle, and why? Describe what an observer sees as time progresses when they are situated farther than R from the origin.

Exercise 6.6 (Equipartition of energy). The *energy* function for the wave equation is

$$E(u(t,\cdot)) = \frac{1}{2}\int_{\mathbb{R}^n} |\partial_t u(t,x)|^2 + |\nabla u(t,x)|^2 \, dx \ .$$

By reason of physical analogy, the first term

$$E_K(u(t,\cdot)) = \frac{1}{2}\int_{\mathbb{R}^n} |\partial_t u(t,x)|^2 \, dx$$

is called the *kinetic energy*, and the second term

$$E_P(u(t,\cdot)) = \frac{1}{2}\int_{\mathbb{R}^n} |\nabla u(t,x)|^2 \, dx$$

is called the *potential energy*. The total energy is a conserved quantity for solutions of the wave equation.

Show that, on average, the kinetic energy and the potential energy for a solution of the wave equation are equal; this is in the sense that

$$\lim_{T\to\infty} \frac{1}{T}\int_0^T E_K(u(t,\cdot))\, dt = \lim_{T\to\infty} \frac{1}{T}\int_0^T E_P(u(t,\cdot))\, dt \ .$$

Hint: You may assume that the initial data (f, g) are in Schwartz class. And you may consider using the solution by Fourier synthesis for this problem and apply the Plancherel identity, that is, the statement that

$$\|f\|_{L^2}^2 = \|\hat{f}\|_{L^2}^2 \ .$$

Projects: Chapter 6

Project 6.7 (Method of descent in $\mathbb{R}^n$ for even dimensions $n = 2m$). The solution of the wave equation in even spatial dimensions is different in character than in odd dimensions. This is described in the case of $\mathbb{R}^2$ in Exercise 6.5 above. A similar procedure of descent gives an explicit spherical means solution formula for solutions to the wave equation in higher even-dimensional Euclidean space $\mathbb{R}^n$, $n = 2m$. Derive a similar version of the spherical means formulae to express the solution. You will note that, similar to the case of dimension $n = 2$, the solution does not satisfy the strong

Huygens' principle. Assume that the initial data is smooth and compactly supported, derive a decay rate for the solution (in the L^∞-norm), and show that this rate is achieved on the light cones supported over the initial data. Show that inside the light cones the solution decays faster, and derive the rate of decay.

Project 6.8 (The Cauchy problem for the wave equation in the Friedman–Robertson–Walker space-time) [**1**],[**13**]). The metric for a Friedman–Robertson–Walker (FRW) space-time is given in terms of the line element in the form

$$ds^2 = -dt^2 + S^2(t)d\sigma^2 \ , \quad S(0) = 0 \ ,$$

defined on the half space-time $\mathbb{R}^1_+ \times \mathbb{R}^3_x$, where $d\sigma^2 = dx_1^2 + dx_2^2 + dx_3^2$ is the Euclidean metric of each space-like hypersurface $\{(t,x) : t = \text{Const.}\} \simeq \mathbb{R}^3$. Changing time variable

$$\frac{dt}{d\tau} = S(\tau) = \tau^2 \ ,$$

the metric becomes

$$ds^2 = S^2(\tau)(-d\tau^2 + d\sigma^2) \ .$$

This metric describes an emerging space-time from a Big Bang at $t = \tau = 0$. Consider the wave equation on $\mathbb{R}^1_+ \times \mathbb{R}^3_x$ in this metric,

$$\Box u := \frac{1}{S^2}\partial_\tau^2 u + \frac{2\dot{S}}{S^3}\partial_\tau u - \frac{1}{S^2}\Delta_\sigma u = 0 \ .$$

Initial data is given on the Cauchy hypersurface $\{(\tau_0, x)\} \simeq \mathbb{R}^3$,

$$u(\tau_0, x) = g(x) \ , \quad \partial_\tau u(\tau_0, x) = h(x) \ .$$

(a) Making the change of variables

$$v(\tau, x) = \frac{1}{\tau}\partial_\tau(\tau^3 u) \ ,$$

show that $v(\tau, x)$ satisfies the usual wave equation in Minkowski space:

$$\partial_\tau^2 v = \Delta_\sigma v \ , \quad \tau > 0 \ .$$

Show that the initial data for v at $\tau = \tau_0 > 0$ is given by

$$\begin{cases} v(\tau_0, x) = 3\tau_0 g(x) + \tau_0^2 h(x) := \phi(x) \ , \\ \partial_\tau v(\tau_0, x) = 3g(x) + \tau_0^2 \Delta g(x) + \tau_0 h(x) := \psi(x) \ . \end{cases}$$

Thus the solution can be given in terms of spherical means; state this expression for the solution.

(b) The inverse of the transformation is given by

$$\tau^3 u(\tau, x) = \int_0^{\tau - \tau_0} (r + \tau_0) v(r + \tau_0, x)\, dr + \tau_0^3 g(x) \ .$$

Assume (for simplicity) that $h(x) = 0$. Give an expression for $u(\tau, x)$ in terms of $g(x)$ using the spherical means expression for v and the above inverse.

(c) The above expression is for fixed $\tau_0 > 0$ defining the Cauchy surface, and it gives the solution at time $\tau > 0$. Now consider the solution expression at a fixed time $\tau > \tau_0$, and take the limit as $\tau_0 \to 0$. What do you get? Can waves pass through a Big Bang singularity?

(d) Make a sketch of the light cone structure of this problem in the original variables (t, x).

Chapter 7

Dispersion

Schrödinger's equation: this image needs no caption.

The phenomenon of dispersion for a PDE refers to the tendency of solutions to spread out in space as time evolves. The most important cases of this phenomenon occur when components of a solution with different spatial frequencies travel at different velocities. Light passing through a medium such as a raindrop or the glass of a prism is an example of this, with results that give rise to a rainbow (from multiple internal reflections in a cluster of raindrops) or the spectral decomposition of a light beam through a prism.

Other dispersive phenomena in physics include wave propagation in waveguides, ocean waves, and solutions of the Schrödinger equations. We will start this chapter with the latter.

7.1. Schrödinger's equation

The most common case of an evolution equation for which dispersion plays a role is the Schrödinger equation

$$i\partial_t \psi = -\frac{1}{2}\Delta\psi , \qquad \psi : \mathbb{R}^1_t \times \mathbb{R}^n_x \to \mathbb{C} , \tag{7.1}$$

with prescribed initial data

$$\psi(0, x) = \psi_0(x) \in L^2(\mathbb{R}^n) .$$

In the quantum mechanical context, the solution $\psi(t, x)$ is known as the *wave function* or the *complex probability amplitude.* The solution of (7.1) is straightforward via the Fourier transform,

$$\hat{\psi}(t, \xi) = (\mathcal{F}\psi)(t, \xi) = \frac{1}{\sqrt{2\pi}^n} \int e^{-i\xi\cdot x}\psi(t, x)\, dx ,$$

which then must satisfy the parameter family of ODEs

$$i\partial_t \hat{\psi} = \frac{1}{2}|\xi|^2\hat{\psi} , \qquad \hat{\psi}(0, \xi) = \hat{\psi}_0(\xi) .$$

Thus

$$\hat{\psi}(t, \xi) = e^{-\frac{i}{2}|\xi|^2 t}\hat{\psi}_0(\xi) .$$

Reconstituting the wave function with the inverse Fourier transform gives the following expression of the solution:

$$\begin{aligned} \psi(t, x) &= \frac{1}{\sqrt{2\pi}^n} \int e^{i\xi\cdot x} e^{-\frac{i}{2}|\xi|^2 t}\hat{\psi}_0(\xi)\, d\xi \\ &= \frac{1}{(2\pi)^n} \iint e^{i\xi\cdot(x-y)} e^{-\frac{i}{2}|\xi|^2 t}\psi_0(y)\, dy d\xi . \end{aligned} \tag{7.2}$$

Proposition 7.1. *Evaluation of the complex Gaussian integral in* (7.2) *gives the expression*

$$\frac{1}{(2\pi)^n} \int e^{i\xi\cdot(x-y)} e^{-\frac{i}{2}|\xi|^2 t}\, d\xi = \frac{1}{\sqrt{2\pi i t}^n} e^{i\frac{|x-y|^2}{2t}} .$$

Thus the Fourier representation of the solution leads to the expression for the solution involving the Schrödinger kernel

$$\begin{aligned} \psi(t, x) &= \int_{\mathbb{R}^n} \frac{1}{\sqrt{2\pi i t}^n} e^{i\frac{|x-y|^2}{2t}} \psi_0(y)\, dy \\ &:= \int_{\mathbb{R}^n} S(t, x-y)\psi_0(y)\, dy = \mathbf{S}(t)\psi_0(x) . \end{aligned} \tag{7.3}$$

One notices formal similarities of the Schrödinger kernel with the heat kernel and its analogous representation of solutions of the heat equation. On the other hand, the presence of the complex unit i in the exponent radically changes the character of the Schrödinger kernel and the behavior of solutions. Indeed, the heat kernel satisfies

$$H(t,x) = \frac{1}{\sqrt{2\pi t}^n} e^{-\frac{|x|^2}{2t}} \in \mathcal{S} \quad \forall t > 0 \ ,$$

and in particular for $t > 0$, the heat kernel $H(t,x)$ decays very rapidly as $|x| \to +\infty$ (faster than exponential decay). On the other hand,

$$S(t,x) = \frac{1}{\sqrt{2\pi i t}^n} e^{\frac{i|x|^2}{2t}} \in C^\infty \quad \forall t \in \mathbb{R} \ ,$$

but it is not Schwartz class as it does not decay as $|x| \to +\infty$, rather it oscillates rapidly.

Theorem 7.2. (i) *For initial data $\psi_0(x) \in \mathcal{S}$, the solution expression* (7.3) *gives rise to an absolutely convergent integral, and for all $t \in \mathbb{R}^1$, $\psi(t,x) \in \mathcal{S}$.*
(ii) *For initial data $\psi_0(x) \in L^2(\mathbb{R}^n)$, the expression* (7.3) *for the solution converges in the L^2 sense, and for all $t \in \mathbb{R}^1$, $\psi(t,x) \in L^2(\mathbb{R}^n)$.*

Proof. (i) The solution formula (7.3) is that

$$\psi(t,x) = \int_{\mathbb{R}^n} S(t,x-y)\psi_0(y)\,dy = \mathcal{F}^{-1}\big(e^{-\frac{i}{2}|\xi|^2 t}\hat{\psi}_0(\xi)\big) \ ,$$

which is the inverse Fourier transform of a Schwartz class function.
(ii) Again the solution formula gives us that

$$\begin{aligned} \|\psi(t,\cdot)\|_{L^2}^2 &= \int |\hat{\psi}(t,\xi)|^2\,d\xi = \int |e^{-\frac{i}{2}|\xi|^2 t}\hat{\psi}_0(\xi)|^2\,d\xi \\ &= \int |\hat{\psi}_0(\xi)|^2\,d\xi = \|\psi_0\|_{L^2}^2 \ . \end{aligned} \tag{7.4}$$

□

This last fact, that the L^2-norm is preserved by the Schrödinger flow, is central to the interpretation of the wave function in quantum mechanics. Consider initial data $\psi_0(x) \in L^2(\mathbb{R}^n)$, normalized to satisfy $\|\psi_0\|_{L^2} = 1$; this can be viewed as a rescaling of ψ_0 to lie on the unit sphere $\mathbb{S}_1^\infty \subseteq L^2$. Then the quantity

$$|\psi(t,x)|^2\,dx := dP_t(x)$$

is a time-dependent distribution on $x \in \mathbb{R}^n$ with initial distribution

$$|\psi_0(x)|^2\,dx = dP_0(x) \ .$$

The normalization of the wave function implies that

$$P_t(\mathbb{R}^n) = \int_{\mathbb{R}^n} |\psi(t,x)|^2\, dx = \int_{\mathbb{R}^n} |\psi_0(x)|^2\, dx = 1 \ .$$

Theorem 7.2(ii) is the fact that the Schrödinger flow preserves probability, and therefore both $dP_0(x)$ and $dP_t(x)$ are probability distributions. The interpretation of the wave function is that the probability of a quantum particle being in a (measurable) set A at time t is given by

$$P_t(A) = \int_A dP_t(x) = \int_A |\psi(t,x)|^2\, dx \ .$$

Comparing this with the heat equation, where the probability measure in question is $dP_t(x) = u(t,x)\, dx$ (which is a measure when $u_0(x) \geq 0$), under heat flow this sign condition and the L^1-norm of the solution are preserved.

Conservation laws. There are other conservation laws for solutions of Schrödinger's equations. The *kinetic energy* represented by the wave function is given by

$$E(\psi) := \int_{\mathbb{R}^n} \frac{1}{2}|\nabla\psi|^2\, dx \ .$$

Expressing this in Fourier transform variables and using the Plancherel identity, it is straightforward to verify that solutions conserve (kinetic) energy (noting that no potential energy term is included in equation (7.1)):

$$\begin{aligned} E(\psi(t,\cdot)) &= \int_{\mathbb{R}^n} \frac{1}{2}|\nabla\psi(t,x)|^2\, dx = \int_{\mathbb{R}^n} \frac{1}{2}|\xi\hat{\psi}(t,\xi)|^2\, d\xi \\ &= \int_{\mathbb{R}^n} \frac{1}{2}\left|\xi e^{-\frac{i}{2}|\xi|^2 t}\hat{\psi}_0(\xi)\right|^2 d\xi = E(\psi_0) \ . \end{aligned}$$

The probability distribution $dP_t(x)$ is called the *position distribution.* One can also study the *momentum distribution*

$$d\hat{P}_t(\xi) := |\hat{\psi}(t,\xi)|^2\, d\xi \ ,$$

which has initial distribution $d\hat{P}_0(\xi) = |\hat{\psi}_0(\xi)|^2\, d\xi$. Again from the Plancherel identity,

$$\int_{\mathbb{R}^n} d\hat{P}_t(\xi) = \int_{\mathbb{R}^n} |\hat{\psi}(t,\xi)|^2\, d\xi = \int_{\mathbb{R}^n} |\psi(t,x)|^2\, dx = \int_{\mathbb{R}^n} dP_0(x) \ ;$$

thus the total momentum probability, or the zeroth moment of $d\hat{P}_t(\xi)$, is preserved under Schrödinger evolution. Furthermore, all moments of the momentum distribution are preserved.

Theorem 7.3. *Define a Fourier multiplier operator $m(D)$ by the formula*

$$m(D)\psi(x) = \frac{1}{\sqrt{2\pi}^n} \int_{\mathbb{R}^n} e^{i\xi\cdot x} m(\xi)\hat{\psi}(\xi)\, d\xi \ .$$

All such operators acting on a solution of Schrödinger's equation (7.1) satisfy

$$\int_{\mathbb{R}^n} |m(D)\psi(t,x)|^2\,dx = \int_{\mathbb{R}^n} |m(D)\psi_0(x)|^2\,dx\ .$$

Moments of the momentum distribution $d\hat{P}_t(\xi)$ are of this form. Therefore all moments of the momentum distribution are preserved under free Schrödinger flow.

Proof. The result follows from a calculation of the integrals in the statement of the theorem,

$$\begin{aligned}\int |m(D)\psi(t,x)|^2\,dx &= \int |m(\xi)\hat{\psi}(t,\xi)|^2\,d\xi \\ &= \int |m(\xi)\hat{\psi}_0(\xi)|^2\,d\xi = \int |m(D)\psi_0(x)|^2\,dx\ .\end{aligned}$$

□

All moments of the measure $d\hat{P}_t(x)$ are of this form:

$$\hat{m}_k(\psi(t,\cdot)) = \int \xi^k d\hat{P}_t(\xi) = \int \xi^k |\hat{\psi}(\xi)|^2\,d\xi = \int \xi^k d\hat{P}_0(\xi) = \hat{m}_k(\psi_0(\cdot))\ .$$

In particular, it follows that for $k = 2s$, these are Sobolev estimates of the solution

$$\begin{aligned}\hat{m}_{2s}(\psi(t,\cdot)) &= \int |\xi^s\psi(t,\xi)|^2\,d\xi = \int |\partial_x^s\psi(t,x)|^2\,dx \\ &= \int |\partial_x^s\psi_0(x)|^2\,dx = \hat{m}_{2s}(\psi_0)\ .\end{aligned}$$

This leads to a natural question as to how the spatial moments of solutions of the Schrödinger equation behave. These are

$$m_k(\psi(t,x)) = \int_{\mathbb{R}^n} x^k dP_t(x) = \int_{\mathbb{R}^n} x^k |\psi(t,x)|^2\,dx\ .$$

It turns out that for $\psi_0 \in L^2(\mathbb{R}^n)$, the quantities $m_k(\psi(t,\cdot))$ are not necessarily bounded; see Projects 7.7 and 7.8 in this chapter. This is in contrast to the recursive control of the moments of the analogous probability measure given by the flow of the heat equation.

A second proof of the conservation law (7.4) for $\|\psi(t,\cdot)\|_{L^2}^2$ goes as follows. Take a domain $A \subseteq \mathbb{R}^n$, that is, an open subset with sufficiently smooth boundary. Then

$$\begin{aligned}\frac{d}{dt}\int_A \bar{\psi}\psi\,dx &= \int_A \overline{\partial_t\psi}\psi + \overline{\psi}\partial_t\psi\,dx = \int_A \overline{\frac{1}{2i}\Delta\psi}\psi + \overline{\psi}\frac{1}{2i}\Delta\psi\,dx \\ &= -\frac{1}{2i}\int_A \overline{\Delta\psi}\psi - \overline{\psi}\Delta\psi\,dx = \frac{1}{2i}\int_A \overline{\nabla\psi}\cdot\nabla\psi - \overline{\nabla\psi}\cdot\nabla\psi\,dx \\ &\quad + \frac{1}{2i}\int_{\partial A}\big(-\psi\overline{\nabla\psi}\cdot N + \overline{\psi}\nabla\psi\cdot N\big)\,dS_A\ .\end{aligned}$$

This calculation identifies the quantum mechanical flux $F(\psi)$ across the boundary ∂A as

$$F\cdot N = \frac{1}{2i}\big(\psi\overline{\nabla\psi} - \overline{\psi}\nabla\psi\big)\cdot N\ .$$

Now let $A = \mathbb{R}^n$. We find that

$$\frac{d}{dt}\|\psi(t,\cdot)\|_{L^2}^2 = 0\ ;$$

hence $\|\psi(t,\cdot)\|_{L^2}^2 = \|\psi_0\|_{L^2}^2$ for all $t\in\mathbb{R}_t^1$. The advantage to this proof is that it is very general, and in particular it applies to the Schrödinger equation with potential

$$i\partial_t\psi = -\frac{1}{2}\Delta\psi + V(x)\psi\ , \tag{7.5}$$

where $V(x)$ represents the interaction of the quantum particle in question with a background potential energy. Two examples that we have already discussed in Chapter 2 are the quantum harmonic oscillator and the hydrogen atom. The potential for the first is given by

$$V(x) = \frac{1}{2}|x|^2\ ,$$

for which solutions of the associated Schrödinger equation (7.5) correspond to the dynamics of a quantum particle trapped in a confining potential. For the second example, let $x\in\mathbb{R}^3$. The potential energy of the nucleus of the hydrogen atom is given by

$$V(x) = -\frac{Z}{|x|}\ ,$$

where the associated wave function $\psi(t,x)$ describes the dynamics of the electron of the hydrogen atom, which is a linear superposition of electron orbitals and extended, or radiating, states.

7.2. Heisenberg uncertainty principle

Returning to the discussion of moments of a function $\psi(x)\in L^2$, we have defined the kth spatial moments and momentum space moments as

$$m_k(\psi_0) = \int x^k|\psi_0(x)|^2\,dx\ , \qquad \hat{m}_k(\psi_0) = \int \xi^k|\hat{\psi}_0(\xi)|^2\,d\xi\ .$$

We have furthermore shown that the moments $m_0(\psi(t,\cdot))$ and $\hat{m}_k(\psi(t,\cdot))$ are constants of Schrödinger evolution. Introducing the Dirac *"bra"* $:= \langle$ and *"ket"* $:= \rangle$ notation for inner products, the first moments of position and momentum of an arbitrary wave function $\psi(x)$ can be written as

$$m_1(\psi) = \int_{\mathbb{R}^n} x|\psi(x)|^2\,dx = \langle\psi|x|\psi\rangle\,,$$

$$\hat{m}_1(\psi) = \int \xi|\hat{\psi}(\xi)|^2\,d\xi = \int \overline{\psi}(x)\frac{1}{i}\partial_x\psi(x)\,dx = \langle\psi|\frac{1}{i}\partial_x|\psi\rangle\,.$$

Since we are working in the setting $x \in \mathbb{R}^n$, the operators $\frac{1}{i}\partial_x$ and x are vectorial, and therefore so are the moments $m_1(\psi)$ and $\hat{m}_1(\psi)$.

Proposition 7.4. *Assume that the wave function $\psi(x) \not\equiv 0$ and that $m_1(\psi), \hat{m}_1(\psi) < +\infty$. Under translation and change of phase, $\psi(x)$ may be adjusted so that $m_1(\psi) = 0 = \hat{m}_1(\psi)$.*

Proof. Addressing first $m_1(\psi)$, change variables by translation $x \mapsto x' - x_0$, and use the notation that $\tau_{x_0}(\psi) := \psi(x - x_0)$. Then

$$\begin{aligned} m_1(\psi) &= \int x|\psi(x)|^2\,dx = \int (x' - x_0)|\psi(x' - x_0)|^2\,dx' \\ &= m_1(\tau_{x_0}\psi) - x_0\, m_0(\tau_{x_0}\psi)\,. \end{aligned}$$

Since $\psi \not\equiv 0$, then $m_0(\psi) = m_0(\tau_{x_0}\psi) \neq 0$, and thus we may choose

$$x_0 = -\frac{m_1(\psi)}{m_0(\psi)}$$

so that $m_1(\tau_{x_0}\psi) = 0$. Similarly, consider

$$\hat{m}_1(\psi) = \int \xi|\hat{\psi}|^2\,d\xi = \int \overline{\psi}(x)\frac{1}{i}\partial_x\psi(x)\,dx\,.$$

Then with a change of phase,

$$\begin{aligned} \hat{m}_1(e^{i\xi_0\cdot x}\psi) &= \int \overline{e^{i\xi_0\cdot x}\psi(x)}\frac{1}{i}\partial_x\big(e^{i\xi_0\cdot x}\psi(x)\big)\,dx \\ &= \int \overline{\psi}(x)\frac{1}{i}\partial_x\psi(x)\,dx + \xi_0\int \overline{\psi}\psi\,dx \\ &= \hat{m}_1(\psi) + \xi_0\hat{m}_0(\psi)\,. \end{aligned}$$

Again, one may choose ξ_0 so that $\hat{m}_1(e^{i\xi_0\cdot x}\psi) = 0$. □

The *variance* $\Delta_x(\psi)$ of a wave function quantifies its deviation from its mean value, which we can write as follows. Assume that $m_1(\psi) = 0 = \hat{m}_1(\psi)$

without loss of generality due to Proposition 7.4. Then we may write

$$m_2(\psi) = \int x_j x_\ell |\psi(x)|^2\, dx = \langle \psi | x_j x_\ell | \psi \rangle \ ,$$

$$\widehat{m}_2(\psi) = \int \overline{\frac{1}{i}\partial_{x_j}\psi}\ \frac{1}{i}\partial_{x_\ell}\psi\, dx = \int \xi_j \xi_\ell |\hat\psi(\xi)|^2\, d\xi = \langle \hat\psi | \xi_j \xi_\ell | \hat\psi \rangle \ ,$$

$$\Delta_x(\psi) = \mathrm{tr}\ (m_2(\psi))\ , \qquad \Delta_\xi(\psi) = \mathrm{tr}\ (\widehat{m}_2(\psi))\ .$$

Theorem 7.5 (Heisenberg uncertainty principle). *Consider a normalized* $\psi(x) \in B_1(0) \subseteq L^2$ *such that*

$$m_0(\psi) = 1\ , \quad m_1(\psi) = 0\ , \quad \widehat{m}_1(\psi) = 0\ .$$

Then there is a lower bound on the product of the variances; namely,

$$1 = m_0(\psi)^2 \le 4 m_2(\psi)\widehat{m}_2(\psi) = \frac{4}{n^2}\Delta_x(\psi)\Delta_\xi(\psi)\ . \tag{7.6}$$

In somewhat more generality, the above inequality follows from the statement that for any $\psi \in L^2$ and any $1 \le j \le n$,

$$\frac{1}{4}\|\psi\|_{L^2}^4 \le \int x_j^2 |\psi(x)|^2\, dx \int \big|\frac{1}{i}\partial_{x_j}\psi\big|^2\, dx\ . \tag{7.7}$$

Proof. We will prove (7.7). Assume that $\psi \in \mathcal{S}$ so that we may freely take derivatives and moments. Calculate the commutator

$$[\frac{1}{i}\partial_{x_j}, x_j]\psi = \frac{1}{i}\partial_{x_j}(x_j\psi) - x_j\frac{1}{i}\partial_{x_j}\psi = \frac{1}{i}\psi\ .$$

Using the commutator, there is an identity

$$\frac{1}{i}\|\psi\|_{L^2}^2 = \int \bar\psi([\frac{1}{i}\partial_{x_j}, x_j]\psi)\, dx \ = \int \overline{\frac{1}{i}\partial_{x_j}\psi}\, x_j\psi\, dx - \int \overline{x_j\psi}\ \frac{1}{i}\partial_{x_j}\psi\, dx\ .$$

Therefore

$$\int \overline{\psi}([\frac{1}{i}\partial_{x_j}, x_j]\psi)\, dx = 2\,\mathrm{im} \int \overline{\frac{1}{i}\partial_{x_j}\psi}\, x_j\psi\, dx\ ,$$

and using the Cauchy–Schwarz inequality, we have

$$\Big|\int \bar\psi([\frac{1}{i}\partial_{x_j}, x_j]\psi)\, dx\Big| \le 2\Big(\int \big|\frac{1}{i}\partial_{x_j}\psi\big|^2\, dx\Big)^{1/2}\Big(\int x_j^2|\psi(x)|^2\, dx\Big)^{1/2}\ .$$

Therefore we obtain the statement in (7.7) for this ψ,

$$\|\psi\|_{L^2}^2 \le 2\|\partial_{x_j}\psi\|_{L^2}\|x_j\psi\|_{L^2}\ .$$

To obtain the statement of Theorem 7.5 concerning the variances, square this inequality and sum over $1 \le j \le n$. The LHS gives $n^2\|\psi\|_{L^2}^4$ while the RHS gives

$$4\big(\Sigma_{j=1}^n \|\partial_{x_j}\psi\|_{L^2}\|x_j\psi\|_{L^2}\big)^2 \le 4\|\nabla\psi\|_{L^2}^2\|x\psi\|_{L^2}^2\ .$$

This is the desired inequality, for $\psi \in \mathcal{S}$ at least. The general case follows from the fact that $\mathcal{S} \subseteq L^2$ is a dense set. We may thus approach a general

$\psi \in L^2$ by a sequence of Schwartz class approximates in such a way that the LHS of (7.7) converges in the sense of L^2 (the RHS may well diverge if, for example, our target L^2 function is not also H^1). □

7.3. Phase and group velocities

The majority of the problems that we have addressed so far in this course have been for constant coefficient PDEs, which is to say that they are of the form

$$P(D)u = \sum_{|\alpha|\leq m} p_\alpha D^\alpha u = 0$$

with $D = \frac{1}{i}\partial_x$ and $D^\alpha = \prod_{j=1}^{n} D_j^{\alpha_j} = (-i)^\alpha \prod_{j=1}^{n} \partial_{x_j}$. In particular, the evolution equations have been posed as

$$\begin{aligned} \partial_t u &= P(D)u \ , \\ u(x,0) &= u_0(x) \ . \end{aligned} \tag{7.8}$$

Our examples have been the Schrödinger equation

$$P(D)u = \frac{i}{2}(\partial_{x_1}^2 + \cdots + \partial_{x_n}^2) = -\frac{i}{2}|D|^2 \ , \tag{7.9}$$

the heat equation

$$P(D)u = \frac{1}{2}\Delta = -\frac{1}{2}|D|^2 \ , \tag{7.10}$$

and the wave equation, for which we must write (7.8) as a system of PDEs,

$$\partial_t \begin{pmatrix} u \\ p \end{pmatrix} = \begin{pmatrix} 0 & I \\ \Delta & 0 \end{pmatrix} \begin{pmatrix} u \\ p \end{pmatrix} , \tag{7.11}$$

so that

$$P(D) = \begin{pmatrix} 0 & 1 \\ -|D|^2 & 0 \end{pmatrix} .$$

The method that we have used in each case is to take the Fourier transform of equation (7.8),

$$\partial_t \hat{v} = P(\xi)\hat{v}, \tag{7.12}$$

and solve the resulting ODE which depends upon the parameter $\xi \in \mathbb{R}^n$:

$$\hat{v}(\xi, t) = e^{P(\xi)t}\hat{v}_0(\xi). \tag{7.13}$$

Modulo questions of convergence, the solution should then be represented as a Fourier integral

$$v(x,t) = \frac{1}{\sqrt{2\pi}^n} \int e^{i\xi\cdot x} e^{P(\xi)t} \hat{u}(\xi) d\xi \ . \tag{7.14}$$

The wave equation gives a good example of this method; in matrix form,

$$P(\xi) = \begin{pmatrix} 0 & 1 \\ -|\xi|^2 & 0 \end{pmatrix} \text{ and } e^{P(\xi)t} = \begin{pmatrix} \cos(|\xi|t) & \frac{1}{|\xi|}\sin(|\xi|t) \\ -|\xi|\sin(|\xi|t) & \cos(|\xi|t) \end{pmatrix} .$$

Since the above sequence of calculation can be reproduced for any constant coefficient operator $P(D)$, one should guess that the Fourier kernel $e^{P(\xi)t}$ of the solution operator cannot give rise to a well-defined distribution kernel in all cases. There is, however, a simple criterion in the scalar case for guaranteeing that we can proceed in this fashion.

Definition 7.6. Suppose that the polynomial $iP(\xi)$ is real valued for all $\xi \in \mathbb{R}^n$. Then $P(\xi)$ is said to be of *real type*, and

$$\omega(\xi) = iP(\xi)$$

is known as the *dispersion relation.*

The idea of a linear evolution equation with constant coefficients can be extended to more general operators which are defined in terms of Fourier multipliers which are symbols.

Definition 7.7. A Fourier multiplier is *of real type* if $\omega(\xi)$ is real for all $\xi \in \mathbb{R}$. A Fourier multiplier $\omega(D) \in \mathbb{C}^\infty(\mathbb{R}^n \setminus \{0\})$ is called a *symbol* if for some m, for all multi-indices β,

$$|\partial_\xi^\beta \omega(\xi)| \le C_\beta(1 + |\xi|^2)^{(m-|\beta|)/2} .$$

The least value m for which this holds is the *order* of the symbol.

Proposition 7.8. *In the case that $\omega(D)$ is a continuous Fourier multiplier of real type, the initial value problem*

$$\partial_t u = -i\omega(D)u \tag{7.15}$$

is well-posed in $L^2(\mathbb{R}^n)$.

Proof. Well-posedness will mean in this context that if the initial data $u_0(x) \in L^2(\mathbb{R}^n)$, then there exists a solution $u(x,t)$ of (7.15) with the properties that for each $t \in \mathbb{R}$, $u(x,t) \in L^2(\mathbb{R}^n)$, and that $\lim_{t\to 0} \|u(x,t) - u_0(x)\|_{L^2(\mathbb{R}^n)} = 0$, which is to say that the solution is continuous in L^2. To show this, we will use the Fourier integral expression for a solution

$$u(x,t) = \frac{1}{\sqrt{2\pi}^n} \int e^{i\xi\cdot x} e^{-i\omega(\xi)t} \hat{u}_0(\xi)\, d\xi . \tag{7.16}$$

Using again the Plancherel identity, $\|u(x,t)\|_{L^2(\mathbb{R}^n_x)} = \|e^{-i\omega(\xi)t}\hat{u}_0(\xi)\|_{L^2(\mathbb{R}^n_\xi)} = \|\hat{u}_0(\xi)\|_{L^2(\mathbb{R}^n_\xi)} = \|u_0\|_{L^2(\mathbb{R}^n_x)}$. The continuity in L^2 follows arguments that we have also seen before. □

I want to talk in more detail about the character of solutions of equation (7.15), and in particular their long time asymptotic behavior, using an analysis of the Fourier integral representation (7.16) and the dispersion relation $\omega(\xi)$. For this analysis, we ask that $\omega(\xi)$ be a symbol of real type. The Fourier integral represents the solution as a superposition of complex exponentials

$$\begin{aligned} u(x,t) &= \frac{1}{(2\pi)^n} \int\int e^{i(\xi\cdot(x-y)-\omega(\xi)t)} u_0(y)\, dy d\xi \\ &= \frac{1}{(2\pi)^n} \int\int e^{i(\xi\cdot\frac{(x-y)}{t}-\omega(\xi))t} u_0(y)\, dy d\xi \ . \end{aligned}$$

Definition 7.9 (Kelvin). The *phase velocity* of the solution given in equation (7.15) is defined to be

$$\vec{c}_p = \omega(\xi)\frac{\xi}{|\xi|^2} \ .$$

The *group velocity* is given by

$$\vec{c}_g = \partial_\xi \omega(\xi) \ .$$

It is useful to describe the significance of these terms. Consider the complex exponential for some fixed *wave vector* $k \in \mathbb{R}^n$,

$$u(x,t) = e^{i(k\cdot x - \omega(k)t)} = e^{ik\cdot(x-\omega(k)\frac{k}{|k|^2}t)} = e^{i\Theta(t,x;k)} \ ,$$

which is a plane wave solution to (7.15). Inspecting the exponent of this expression, the phase function $\Theta(t,x;k)$ takes a constant value θ on loci of points in space-time which lie on hyperplanes $k\cdot(x - \frac{\omega(k)}{|k|}\frac{k}{|k|}t) = \theta$. That is, the points of constant phase are moving in direction $\frac{k}{|k|}$ with speed $\frac{\omega(k)}{|k|} = |\vec{c}_p(k)|$. For example, taking the real (or imaginary) part of the solution, this shows that the velocity of the peak or the trough of an individual oscillation is given by $c_p(k)$.

Now consider a more distributed set of frequencies in the initial data in the form of a wave packet. This is to say that we take data which decomposes into two features, a complex exponential multiplying a localized function:

$$u_0(x) = e^{ik\cdot x} a_0(x) \ . \tag{7.17}$$

This localized function $a_0(x) \in \mathcal{S}$ represents an envelope which is slowly varying with respect to the oscillation of the complex exponential factor. We will show that this wave packet will propagate with group velocity $\vec{c}_g(k)$. To illustrate the fact that for a particular equation, the phase and group velocities are not generally the same, here are several examples.

(1) The Schrödinger equation. $\omega(\xi) = \frac{1}{2}|\xi|^2$, $c_p = \frac{1}{2}\xi$, $c_g = \partial_\xi\omega = \xi$. The group velocity is twice the phase velocity.

(2) The wave equation. The matrix

$$P(\xi) = \begin{pmatrix} 0 & 1 \\ -|\xi|^2 & 0 \end{pmatrix}$$

has eigenvalues $\omega(\xi) = \pm|\xi|$. The phase and group velocities coincide for each of the two eigenvectors of $P(\xi)$:

$$c_p = \pm\frac{\xi}{|\xi|} = c_g \ .$$

(3) The Airy equation (related to the KdV equation) for $x \in \mathbb{R}^1$ is

$$\partial_t u = \partial_x^3 u \ , \quad P(D) = -iD^3 \ .$$

The dispersion relation $\omega(\xi) = \xi^3$ has

$$c_p = \xi^2 \geq 0 \ , \quad c_g = 3\xi^2 = 3c_p \ .$$

(4) Free surface water waves. The dispersion relation is $\omega(\xi) = \sqrt{g|\xi|}$ for surface waves over an infinitely deep body of water. The parameter g is related to the acceleration of gravity and the density of water. Then

$$c_p = \sqrt{\frac{g}{|\xi|}}\frac{\xi}{|\xi|} \ , \quad c_g = \frac{1}{2}\sqrt{\frac{g}{|\xi|}}\frac{\xi}{|\xi|} \ ,$$

from which we see that c_g is slower than c_p. An astute observer can see this relationship in the behavior of ripples in disturbances in a pond.

7.4. Stationary phase

Starting with wave packet initial data (7.17), a solution of equation (7.8) is expressed in Fourier integrals, giving the familiar expressions

$$(7.18) \qquad \begin{aligned} u(t,x) &= \frac{1}{\sqrt{2\pi}^n}\int e^{i(\xi\cdot(x-y)-\omega(\xi)t)}e^{ik\cdot y}a_0(y)\,dyd\xi \\ &= e^{ik\cdot x}\frac{1}{\sqrt{2\pi}^n}\int e^{i(\xi'\cdot x-\omega(k+\xi')t)}\hat{a}_0(\xi')\,d\xi' \ . \end{aligned}$$

The latter integral follows from the change of variables $\xi = k + \xi'$. The following argument is a common one in formal asympotic analysis, and although it has many aspects which are difficult to make rigorous in general, it provides very useful intuition for the behavior of solutions of dispersive equations. On a formal level, consider ξ' as a perturbation of the wave number k. The Taylor expansion of the phase function is

$$\omega(k+\xi') = \omega(k) + \partial_\xi\omega(k)\cdot\xi' + \frac{1}{2}\langle\xi', \partial_\xi^2\omega(k)\xi'\rangle + \mathcal{O}(|\xi'|^3) \ .$$

In these terms the integral in (7.18) is

$$u(t,x) = e^{i(k\cdot x-\omega(k)t)}\frac{1}{\sqrt{2\pi}^n}\int e^{i(\xi'\cdot x-\partial_\xi\omega(k)\cdot\xi' t)}e^{(\frac{i}{2}\langle\xi',\partial_\xi^2\omega(k)\xi'\rangle+i\mathcal{O}(|\xi'|^3))}\hat{a}_0(\xi')\,d\xi' .$$

The phase function in this integral is

$$\Theta(t,x,\xi;k) = \xi'\cdot[x-\partial_\xi\omega(k)t]+\frac{1}{2}\langle\xi',\partial_\xi^2\omega(k)\xi'\rangle + 1\mathcal{O}(|\xi'|^3) .$$

Consider the integral for a point (t,x) in space-time; its contribution to the solution is negligible for large $|k|$ or for long time $|t|$ if $[x-\partial_\xi\omega(k)t]\neq 0$ (at least it is so under further hypotheses on the integrand). Therefore we consider only (t,x) which satisfy $[x-\partial_\xi\omega(k)t]=0$. For such space-time points, the integral representing the solution $u(t,x)$ is reduced to

$$\frac{1}{\sqrt{2\pi}^n}\int e^{\frac{i}{2}\langle\xi',\partial_\xi^2\omega(k)\xi'\rangle+i\mathcal{O}(|\xi'|^3)}\hat{a}_0(\xi')\,d\xi' = \frac{e^{i\frac{\pi}{4}\operatorname{sgn}\partial_\xi^2\omega(k)}}{\sqrt{|\det\partial_\xi^2\omega(k)|t^n}}\hat{a}_0(0)+o(1) , \tag{7.19}$$

where $\operatorname{sgn}\partial_\xi^2\omega(k)$ is the signature of the Hessian matrix (the signature of a symmetric matrix Λ, denoted by $\operatorname{sgn}\Lambda$, is the number of positive eigenvalues of Λ minus the number of negative ones). Expression (7.19) follows from contour deformation of ξ into the complex plane; it gives a valid asymptotic formula under a number of further assumptions, including the hypothesis that $\det\partial_\xi^2\omega(k)\neq 0$. This asymptotic calculation is discussed in Exercise 7.6. If the hypothesis on the nondegeneracy of the Hessian is not satisfied, the contribution to the asymptotic behavior of the solution can be more prominent as the resulting asymptotic expression decays less rapidly; this plays a role in the Kelvin ship wake problem which is the topic of Project 7.10. In any case, the conclusion is that the principal contribution to the solution $u(t,x)$ is for the space-time points such that

$$\frac{x}{t}=\partial_\xi\omega(k) ,$$

which is to say that the wave packet propagates at the group velocity $\vec{c}_g(k)$.

The second part of this section on the method of stationary phase is dedicated to a more robust analysis of the Fourier integral expression for the solution

$$u(t,x)=\frac{1}{\sqrt{2\pi}^n}\int e^{it(\frac{x\cdot\xi}{t}-\omega(\xi))}\hat{u_0}(\xi)\,d\xi . \tag{7.20}$$

Consider the data $\hat{u_0}(\xi)$ supported in $B_r(\xi_0)$, and take any point $(x,t)\in\mathbb{R}^{n+1}$ such that for $\xi\in\operatorname{supp}(\hat{u_0}(\cdot))$, $|\frac{x}{t}-\partial_\xi\omega(\xi)|>R$. Make use of the identity

$$e^{it(\frac{x\cdot\xi}{t}-\omega(\xi))} = \frac{-i(\frac{x}{t}-\partial_\xi\omega(\xi))}{t|\frac{x}{t}-\partial_\xi\omega(\xi)|^2}\cdot\partial_\xi(e^{it(\frac{x\cdot\xi}{t}-\omega(\xi))})$$

in expression (7.20) to rewrite it as

$$\begin{aligned}(7.21)\qquad u(x,t) &= \frac{1}{\sqrt{2\pi}^n}\int \frac{-i(\frac{x}{t}-\partial_\xi\omega(\xi))}{t|\frac{x}{t}-\partial_\xi\omega(\xi)|^2}\cdot\partial_\xi(e^{it(\frac{x\cdot\xi}{t}-\omega(\xi))})\hat{u_0}(\xi)\,d\xi \\ &= \frac{1}{\sqrt{2\pi}^n}\int e^{it(\frac{x\cdot\xi}{t}-\omega(\xi))}\partial_\xi\cdot(\frac{i(\frac{x}{t}-\partial_\xi\omega(\xi))}{t|\frac{x}{t}-\partial_\xi\omega(\xi)|^2}\hat{u_0}(\xi))\,d\xi\ .\end{aligned}$$

The critical fact is that the denominator does not vanish on the support of the integrand, by the choice of $(x,t)\in\mathbb{R}^{n+1}$. Measuring the absolute value of the solution,

$$\begin{aligned}|u(x,t)| &\le \frac{1}{\sqrt{2\pi}^n}\frac{1}{|t|}\int\Big|\partial_\xi\cdot(\frac{i(\frac{x}{t}-\partial_\xi\omega(\xi))}{t|\frac{x}{t}-\partial_\xi\omega(\xi)|^2}\hat{u_0}(\xi))\Big|\,d\xi \\ &\le \frac{C}{|t|}\int\Big|\frac{1}{|\frac{x}{t}-\partial_\xi\omega(\xi)|}\partial_\xi\hat{u_0}(\xi)+\frac{\partial_\xi^2\omega}{|\frac{x}{t}-\partial_\xi\omega(\xi)|^2}\hat{u_0}(\xi)\Big|\,d\xi \\ &\le \frac{C_1}{R|t|}\ ,\end{aligned}$$

for $R>1$ at least. Thus the solution is shown to decay in time along the ray $\frac{x}{t}=\partial_\xi\omega(\xi)$. Repeating the integration by parts that we performed above,

$$\begin{aligned}|u(x,t)| &= \Big|\frac{1}{\sqrt{2\pi}^n}\int e^{it(\frac{x\cdot\xi}{t}-\omega(\xi))}(i\partial_\xi\cdot\frac{i(\frac{x}{t}-\partial_\xi\omega(\xi))}{t|\frac{x}{t}-\partial_\xi\omega(\xi)|^2})^N\hat{u_0}(\xi)\,d\xi\Big| \\ &\le \frac{C_N}{R^N|t|^N}\end{aligned}$$

for large $|t|$. The picture of this is as follows.

Theorem 7.10. *For initial data $u_0(x)\in\mathcal{S}$ such that $\operatorname{supp}(\hat{u_0}(\xi))\subseteq B_r(\xi_0)$, the solution $u(x,t)$ is also in $\mathcal{S}$ for every t. Furthermore, on the set $B=\{(x,t):\inf_{\xi\in B_r(\xi_0)}|\frac{x}{t}-\partial_\xi\omega(\xi)|>R\}$, we have for all N*

$$|u(x,t)|\le\frac{C_N}{R^N|t|^N}\ .$$

This is the statement that away from the rays of stationary phase $\{\frac{x}{t}-\partial_\xi\omega(\xi)=0;\xi\in\operatorname{supp}(\hat{u_0}(\cdot))\}$, the solution decays rapidly in space and time.

It may seem as though the above result only applies to quite special initial functions $u_0(x)$, but in fact it is rather generally applicable. For example, the regions describing stationary phase asymptotics for two wave packet solutions are illustrated in Figure 1. Indeed, we may furthermore decompose any initial data $u_0(x)$ into a sum of wave packets through a partition of unity.

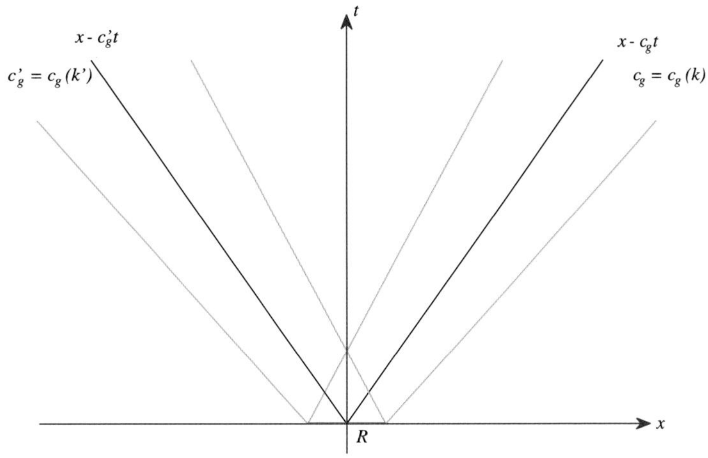

Figure 1. The channels travelled by wave packets with carrier oscillation k and k'. The black lines have slopes $\frac{x}{t} = \partial_\xi(k)$ and $\partial_\xi(k')$, representing propagation of wave packets at their respective group velocities. The surrounding green conical channels are the regions outside of which the wave packets decay at rate $\frac{C_N}{R^N|t|^N}$.

Lemma 7.11. *There exists* $\{\chi_j(y)\}_{j\in\mathbb{Z}^n}$*, a partition of unity of* $\mathbb{R}^n$*, with each* $\chi_j(y) \in \mathbb{C}_0^\infty(\mathbb{R}^n)$ *such that*

$$\operatorname{supp}(\chi_j) \subseteq j + \frac{3}{2}Q\ , \quad \sum_{j\in\mathbb{Z}^n} \chi_j(y) = 1\ ,$$

$$\chi_j(y) = 1 \quad \textit{on} \quad j + \frac{1}{2}Q\ ,$$

where Q *is the unit cube* $\{y \in \mathbb{R}^n; -1 \le y_j \le 1, j = 1, \dots, n\}$.

Proof. This can be constructed from a test function $\tilde{\chi}_0(y) \in \mathbb{C}_0^\infty$ which satisfies $\operatorname{supp}(\tilde{\chi}_0) \subseteq \frac{3}{2}Q$, $\tilde{\chi}_0 \ge 0$, and $\tilde{\chi}_0(y) = 0$ for $y \in Q$. Set $\tilde{\chi}_j(y) = \tilde{\chi}_0(y-j)$ for all $j \in \mathbb{Z}^n$; it is supported in the cube $j + \frac{3}{2}Q$. We finally have to adjust the magnitude. But this is easy as we notice that for every $y \in m + Q$,

$$\sum_m \tilde{\chi}_m(y) = \sum_{|k-m|<2} \tilde{\chi}_k(y) > 0\ .$$

We can simply divide to get

$$\chi_j(y) = \frac{\tilde{\chi}_j(y)}{\sum_k \tilde{\chi}_k(y)}\ .$$

Notice that $\sum_{0<|k-m|<2} \tilde{\chi}_k(y) = 0$ for $|y-m| < \frac{1}{2}$; hence $\chi_m(y) = 1$ identically on $m + \frac{1}{2}Q$. □

Theorem 7.12. *For initial data $u_0 \in L^2(\mathbb{R}^n)$, the solution $u(x,t)$ of (7.8) is in $L^2(\mathbb{R}^n)$. Furthermore the initial data can be decomposed into a superposition of compactly supported L^2 functions $\sum_{j\in\mathbb{Z}^n}(u_0)_j$ where each $(u_0)_j(x) \in L^2(j+\frac{3}{2}Q)$. The resulting solution is a superposition of wave packets $u_{jk}(x,t) \in \mathcal{S}$, each of which is asymptotically localized by Theorem 7.10 around the ray $\{\frac{x}{t} = \partial_\xi\omega(k)\}$.*

Proof. The initial data $u_0(x)$ can be decomposed into a sum of compactly supported L^2 functions using a partition of unity from Lemma 7.11:

$$u_0(x) = \sum_{j\in\mathbb{Z}^n} (u_0)_j(x) = \sum_{j\in\mathbb{Z}^n} (\chi_j u_0)(x) \ .$$

Secondly, we may decompose the Fourier transform $\widehat{(u_0)_j}$, using again a partition of unity from Lemma 7.11, with the result that $\widehat{(u_0)_j}\chi_k(\xi) \in C_0^\infty$. Therefore $\mathcal{F}^{-1}(\widehat{(u_0)_j}\chi_k(\xi)) \in \mathcal{S}$ with Fourier support localized in a cube $\frac{3}{2}Q + k \subseteq \mathbb{R}^n_\xi$. Theorem 7.10 applies to each component, which behaves as a wave packet, asymptotically being very well localized near rays determined by the relevant group velocity $\{\frac{x}{t} = \partial_\xi\omega(j)\}$. □

Exercises: Chapter 7

Exercise 7.1. The Schrödinger equation with a potential is given as

$$i\partial_t\psi = -\frac{1}{2}\Delta\psi + V(x)\psi \ .$$

(i) Show that for a potential $V(x)$ that is bounded, the $L^2(\mathbb{R}^n)$-norm of the solution is conserved by the equation, namely

$$\frac{d}{dt}\int_{\mathbb{R}^n} |\psi(t,x)|^2\, dx = 0 \ .$$

(ii) What is the quantum mechanical flux F in this case?

Exercise 7.2 (Conservation laws)**.** There is an analogy between conservation laws and the Schrödinger equation, with or without a potential term. The density function for the position distribution probability is given by

$$u(t,x) = |\psi(t,x)|^2 \ .$$

Derive an evolution equation for the function $u(t,x)$ of the form of a conservation law, namely

$$\partial_t u + \nabla\cdot F = 0 \ .$$

What is the function F in terms of the wave function $\psi(t,x)$? In this equation, the vector function F plays the role of the quantum mechanical flux.

Exercise 7.3. Consider the initial value problem for Schrödinger's equation for $x \in [0, 2\pi]$, namely

$$i\partial_t \psi = -\frac{1}{2}\partial_x^2 \psi \ ,$$
$$\psi(0, x) = \psi_0(x) \ .$$

(i) With Dirichlet boundary conditions imposed,

$$\psi(t, 0) = 0 = \psi(t, 2\pi) \ ,$$

and using the method of separation of variables, give an expression for the general solution of the initial value problem. What are the eigenvalues λ_k^D of the Schrödinger operator in this case?

(ii) For periodic boundary conditions

$$\psi(t, x) = \psi(t, x + 2\pi) \ ,$$

give an expression via separation of variables for the general solution of the initial value problem. What are the eigenvalues λ_k^p of the Schrödinger operator in this case?

Exercise 7.4 (Separation of variables)**.** For $x \in \mathbb{R}^2$, suppose that the potential $V(x)$ is a square-well, namely

$$V(x) = 0 \quad \text{for} \quad |x| \geq \frac{1}{2} \ , \qquad V(x) = -E \quad \text{for} \quad -\frac{1}{2} < x < \frac{1}{2} \ .$$

Writing $\psi(x) = v(t)w(x)$, find all energies λ such that $w(x)$ is bounded in $x \in \mathbb{R}^1$.

If $w(x) \to 0$ as $x \to \pm\infty$ in such a way that $\|w\|_{L^2(\mathbb{R})} < +\infty$, then it is called a *bound state* and the associated λ is an eigenvalue of the Schrödinger operator with this potential. How many eigenvalues and bound states does this Schrödinger operator have, and how does this number depend upon E? Express the time evolution of the solution whose initial data is this bound state.

If $w(x)$ is bounded but is not in $L^2(\mathbb{R}^1)$, then it is called an extended state. Discuss the set of extended states of the Schrödinger operator with this potential.

Exercise 7.5. In one space dimension, the Heisenberg uncertainty principle states that

$$\frac{1}{4}\|\psi\|_{L^2}^4 \leq \Big(\int_{\mathbb{R}^1} x^2 |\psi(x)|^2 \, dx\Big)\Big(\int_{\mathbb{R}^1} \Big|\frac{1}{i}\partial_x \psi(x)\Big|^2 \, dx\Big) \ .$$

Is this inequality an equality for some function or functions $\psi(x)$? If so, which function(s) achieves equality?

Exercise 7.6 (Method of stationary phase). In case $\hat{a}(\zeta)$ is entire of exponential type, reduce the one-dimensional integral

$$\frac{1}{\sqrt{2\pi}} \int_{-\infty}^{+\infty} e^{i\frac{\omega''}{2}\xi^2 t} \hat{a}(\xi)\, d\xi = \frac{1}{\sqrt{2\pi}} \int_{\Gamma} e^{i\frac{\omega''}{2}\zeta^2 t} \hat{a}(\zeta)\, d\zeta \ ,$$

where ω'' is real and Γ is a contour in the complex ζ plane consisting of two rays $\Gamma_1 \cup \Gamma_2 = \{re^{i\varphi}\} \cup \{re^{i(\pi+\varphi)}\}$. The angle $\varphi = \mathrm{sgn}\,(\omega'')\pi/4$ is positive or negative depending upon the sign of ω''. The sign is chosen so that the exponential factor is decaying along the rays in question. Furthermore, show that the latter integral is the sum of two terms of similar form, the first being

$$\frac{1}{\sqrt{2\pi}} \int_{\Gamma_1} e^{-\frac{\omega''}{2} r^2 t} \hat{a}(re^{i\varphi})\, e^{i\varphi} dr = \frac{e^{i\frac{\pi}{4}\mathrm{sgn}\,(\omega'')}}{\sqrt{2\pi|\omega''|t}} \int_0^{+\infty} e^{-\frac{\rho^2}{2}} \hat{a}\Big(\frac{\rho e^{i\varphi}}{\sqrt{|\omega''|t}}\Big)\, d\rho \ .$$

The second term is similar.

Now show that if $\hat{a}(\zeta)$ is entire of exponential type, for $\zeta \in \mathbb{C}^n$, the multidimensional complex Gaussian integral asymptotically takes the form (7.19) for large time $|t|$.

Projects: Chapter 7

Project 7.7 (Gaussian wave packets for the Schrödinger equation). Give an explicit solution to the Schrödinger equation

$$i\partial_t \psi = -\frac{1}{2}\Delta\psi$$

for all $t \in \mathbb{R}^1$, with initial data given by

$$\psi_0(x) = a(0) e^{ik_0 \cdot x} e^{-\frac{1}{2}\langle x, Ax\rangle} \ ,$$

where $A = A^T > 0$ is a Hermitian matrix. What are the moments $m_1(\psi_0)$ and $m_1(\psi(t,\cdot))$, as well as $m_2(\psi_0)$ and $m_2(\psi(t,\cdot))$? From these quantities you can calculate $\Delta_x(\psi_0)$ and $\Delta_x(\psi(t,\cdot))$. What are $\Delta_\xi(\psi_0)$ and $\Delta_\xi(\psi(t,\cdot))$?

Project 7.8 (Bounds on moments of the Schrödinger equation). Recall that we have set

$$m_j(\psi) = \int_{\mathbb{R}^1} x^j |\psi(x)|^2\, dx \ , \quad \widehat{m}_\ell(\psi) = \int_{\mathbb{R}^1} \xi^\ell |\hat{\psi}(\xi)|^2\, d\xi \ .$$

For solutions $\psi(t,x)$ of the Schrödinger equation (7.1) with initial data $\psi_0(x)$, we have shown that $m_0(\psi(t,\cdot)) = m_0(\psi_0) = \widehat{m}_0(\psi(t,\cdot))$; this is stated in (7.4).

(i) Show that for $j \geq 1$, there do not exist constants C_j such that

$$|m_j(\psi(t,\cdot))| \leq C_j |m_j(\psi_0)| \ .$$

Hint: Use the solutions derived in Project 7.7.

(ii) On the other hand, show that for all indices ℓ,

$$\widehat{m}_\ell(\psi(t,\cdot)) = \widehat{m}_\ell(\psi_0)\ .$$

(iii) However, for $j = 2p$ even, show that there is a constant C_{2p} such that

$$m_{2p}(\psi(t,\cdot)) \leq C_{2p}\Big(m_{2p}(\psi_0) + \widehat{m}_{2p}(\psi_0) + m_0(\psi_0)\Big)\ .$$

Compare this situation with what is true for $t > 0$ in the case of higher moments of the heat equation.

Project 7.9 (Electron orbitals for the hydrogen atom)**.** In Project 2.16 of Chapter 2 we studied the orbitals of electrons in the quantum mechanical potential of the hydrogen atom nucleus. In that project we derived the equations for the components of the orbitals using the method of separation of variables. In this project we will complete the picture by solving these equations for the special functions that arise from the method. Recall that the dynamics of the wave function for an electron in the potential of a point-like nucleus are given by the Schrödinger equation

$$i\partial_t\psi = -\frac{1}{2}\partial_x^2\psi - \frac{Z}{|x|}\psi\ .$$

The coefficient Z is related to the atomic number of the nucleus and the charge of the electron. The Ansatz for a wave function from the method of separation of variables is that

$$\psi(t,x) = e^{-i\lambda t}\varphi(x) = e^{-i\lambda t}R(r)\Phi(\phi)\Theta(\theta)\ ,$$

where (r,ϕ,θ) are spherical polar coordinates. The three functions $\Phi(\phi)$, $\Theta(\theta)$, and $R(r)$ satisfy the following three equations. Firstly, the dependence of the wave function upon the longitudinal angle $\phi \in \mathbb{T}^1$ is given by

$$\frac{d^2\Phi}{d\phi^2} = -m^2\Phi\ , \quad m \in \mathbb{Z}\ ,$$

for which the solution is evidently $\Phi(\phi) = e^{im\phi}$ for some $m \in \mathbb{Z}$. Secondly, dependence on the azimuthal angle is given by solutions of the *associated Legendre equation*

$$-\frac{1}{\sin(\theta)}\frac{d}{d\theta}\Big(\sin(\theta)\frac{d}{d\theta}\Theta\Big) + \frac{m^2}{\sin^2(\theta)}\Theta = k\Theta\ . \tag{7.22}$$

Finally, the radial dependence of the wave function is given through the *associated Laguerre equation*,

$$\frac{1}{2r^2}\frac{d}{dr}\Big(r^2\frac{d}{dr}R\Big) + (\lambda + \frac{Z}{|x|})R = \frac{k}{2r^2}R\ . \tag{7.23}$$

The goal of this project is to describe the solutions $\Theta(\theta)$ and $R(r)$ of the latter two equations.

(i) (Associated Legendre functions). Equation (7.22) is defined on the domain $-\pi/2 < \theta < +\pi/2$.

(a) Change variables by $\xi = \cos(\theta) \in [-1, 1]$ so that (7.22) is written

$$-\frac{d}{d\xi}\Big((1-\xi^2)\frac{d}{d\xi}\Theta(\xi)\Big) + \frac{m^2}{(1-\xi^2)}\Theta(\xi) = k\Theta(\xi) \ .$$

The points $\xi = \pm 1$ are singular points of the first kind, for which the second term of the LHS is a Weyl limit point case. This is the Euler–Lagrange equation for the Lagrangian functional

$$L(\Theta) = \frac{1}{2}\int_{-1}^{+1} (1-\xi^2)\Big(\frac{d\Theta}{d\xi}\Big)^2 + \frac{m^2}{(1-\xi^2)}\Theta^2(\xi)\, d\xi$$

with the additional constraint that $I(\Theta)$ is held to a constant, where

$$I(\Theta) = \frac{1}{2}\int_{-1}^{+1} \Theta^2(\xi)\, d\xi \ .$$

Namely, the equations can be expressed as

$$\delta L = k\delta I$$

with Lagrange multiplier k.

(b) Make the substitution $\Theta(\xi) = (1-\xi^2)^{|m|/2}F(\xi)$ in the Lagrangian L and in I, with the result that

$$L = \frac{1}{2}\int_{-1}^{+1} (1-\xi^2)^{|m|+1}\Big(\frac{dF}{d\xi}\Big)^2 + |m|(|m|+1)(1-\xi^2)^{|m|}F^2\, d\xi \ ,$$

$$I = \frac{1}{2}\int_{-1}^{+1} (1-\xi^2)^{|m|}F^2\, d\xi \ .$$

Now derive the new Euler–Lagrange equations for the functions $F(\xi)$.

(c) Consider power series expressions for F,

$$F(\xi) = \sum_{j\le 0} f_j \xi^j \ .$$

In order to satisfy the Euler–Lagrange equations, the coefficients must satisfy a recursion relation

$$(j+2)(j+1)f_{j+2} = \big(k - (|m|+j+1)(|m|+j)\big)f_j \ .$$

While in general the series cannot be expected to converge, if $k = \ell(\ell+1)$ for some $\ell > |m|$, the recursion terminates. The resulting polynomial $F_\ell^m(\xi)$ is an associated Legendre polynomial, and we recover that

$$\Theta_\ell^m(\theta) = \sin^{|m|}(\theta)F_\ell^m(\cos(\theta)) \ .$$

The $F_\ell^m(\xi)$ are the orthogonal polynomials over the interval $\xi \in (-1, 1)$ for the weight function $(1-\xi^2)^{|m|}$.

(ii) (Associated Laguerre functions). Equation (7.23) is defined for $r \in \mathbb{R}^1_+$.

(a) Rescale the independent variable in (7.23) by

$$\rho = 2\beta r \ .$$

Take $k = \ell(\ell+1)$ to follow part (i), defining β as follows and defining γ in terms of β:

$$\frac{\lambda}{2\beta^2} = -\frac{1}{4} \ , \qquad \gamma = \frac{Z}{\beta} \ .$$

The resulting equation is now expressed as

$$\frac{1}{\rho^2}\frac{d}{d\rho}\Big(\rho^2\frac{dR}{d\rho}\Big) + \Big(-\frac{1}{4} - \frac{\ell(\ell+1)}{\rho^2} + \frac{\gamma}{\rho}\Big)R = 0 \ .$$

(b) For large ρ, bounded solutions behave as $R(\rho) \sim e^{-\rho/2}$ which motivates the Ansatz

$$R(\rho) = e^{-\rho/2}\rho^\ell G(\rho) \ .$$

The Lagrangian for this problem is

$$L = \int_0^{+\infty} \frac{1}{2}\Big(\frac{dR}{d\rho}\Big)^2 \rho^2 \, d\rho + \int_0^{+\infty} \frac{1}{2}\Big(-\frac{1}{4} - \frac{\ell(\ell+1)}{\rho^2} + \frac{\gamma}{\rho}\Big)R^2 \, d\rho \ .$$

Substitute the Ansatz into the Lagrangian and derive the Euler–Lagrange equations $\delta L = 0$:

$$\frac{d^2G}{d\rho^2} + \Big(\frac{2\ell}{\rho} - 1\Big)\frac{dG}{d\rho} + \Big(\frac{\gamma-\ell}{\rho} - \frac{2\ell}{\rho^2}\Big)G = 0 \ .$$

(c) As with the case of the Legendre polynomials, substitute a formal power series for the solution

$$G(\rho) = \sum_{j\geq 0} g_j \rho^j \ .$$

In order to satisfy the Euler–Lagrange equations, the coefficients must satisfy the recursion relation

$$[(j+2)(j+1) + 2\ell(j+1)]g_{j+2} = (j+1+\ell-\gamma)g_{j+1} \ .$$

The series terminates when $n = \gamma \geq (\ell+1)$ is an integer, giving an associated Laguerre polynomial $G_{n\ell}(\rho)$ as a solution.

Following our rescaling of the problem, $\gamma = n$; therefore

$$\beta = \frac{Z}{n} \ , \qquad \lambda = \lambda_n = -\frac{2\beta^2}{4} = -\frac{Z^2}{2n^2} \ .$$

These λ_n are the $L^2(\mathbb{R}^3)$ eigenvalues of the hydrogen atom, while the radial and angular dependence of the eigenfunctions are determined by the above products of associated Laguerre functions $R_{n\ell}(r)$, associated Legendre functions $\Theta_\ell^m(\theta)$, and finally the complex exponentials $\Phi_m(\phi)$. The possible quantum numbers of this eigenstate are determined by the constraints on the indices described above, namely

$$\ell + 1 \leq n \ , \quad |m| \leq \ell \ .$$

The lowest energy electron state λ_1 is for $n = 1$, where only $\ell, m = 0$ are permitted. The next energy level is λ_2, which corresponds to excited states of electrons in the next energy shell. There are three excited states that correspond to orbitals whose angular dependence is given by the spherical harmonics $Y_1^m(\phi, \theta) = \Phi_m(\phi)\Theta_1^m(\theta)$ for $m = -1, 0, 1$ and whose radial dependence is given by $R_{21}(r)$, along with a fourth independent eigenstate $R_{20}(r)Y_0^0$ which is purely radial. The well-known fact that the electron shells contain double these amounts of electrons is due to the extra spin degree of freedom, which doubles the number of possible electrons in a shell to $2, 8 \cdots$. When the spectral parameter λ is positive, this problem has continuous spectrum and extended states, solutions which are not in $L^2(\mathbb{R}^3)$. It corresponds to the presence of electrons that are not bound to the nucleus of the hydrogen atom.

Project 7.10 (Kelvin ship wake problem)**.** This problem concerns the shape of the waves in the wake of a ship. It is a famous calculation of Thomson (Lord Kelvin) [**11**] that explains that when moving in a straight line, the waves in a ship's wake form a vee behind the ship of universal opening angle that is independent of the speed at which the ship is moving. This calculation is at the heart of the origin of the method of stationary phase.

The equations of water waves. The Euler equations of fluid mechanics are a system of equations articulated in terms of the velocity $u(t, x)$ of the fluid. In the case of water waves, it also involves the dynamics of the upper *free surface* which defines the moving fluid domain. We consider a fluid which is infinitely deep, and we use the linear equations of motion, which is a valid hypothesis for small deformations of the free surface and small fluid velocities. We therefore assume that the fluid domain is $\mathbb{R}^3_- = \{(x_1, x_2, y) \in \mathbb{R}^3 \, , y < 0\}$. In this domain, the fluid vector field is incompressible and irrotational, namely

$$\nabla \cdot u = 0 \ , \quad \nabla \times u = 0 \ ,$$

which implies that there exists a velocity potential function $\varphi(t, x, y)$ defined in $\mathbb{R}^3_-$ and satisfying

$$u = \nabla \varphi \ , \quad \Delta \varphi = 0 \ .$$

For boundary conditions for φ, we assume that $\nabla \varphi \to 0$ as $y \to -\infty$, while on the free surface we have

$$\begin{aligned} \partial_t \eta &= \partial_y \varphi \ , \\ \partial_t \varphi &= -g\eta + F \ . \end{aligned} \tag{7.24}$$

The first equation expresses the fact that $\eta(t,x)$ moves vertically according to the vertical component of the fluid velocity. The second equation expresses Newton's law of forces; here g is the acceleration of the restoring force of gravity, and in the absence of other forces the value of the pressure on the free surface is taken to be zero. The function F describes any other additional forces that impinge upon the upper free surface.

(i) Equations (7.24) describe the evolution of the free surface in terms of the variables $\eta(t,x)$ and $\varphi(t,x,y)$, for which the only reference to the motion of the body of the fluid is through the term $\partial_y \varphi(t,x,0)$, where $\varphi(t,x,y)$ is a harmonic function in the fluid domain. Conveniently, this is the quantity expressed by the Dirichlet–Neumann operator for the fluid domain, which in the case of $\mathbb{R}^n_+$ is given in formula (4.38). Denoting the surface boundary values of $\varphi(x,y)$ by $\xi(x) = \varphi(x,0)$, show that the Dirichlet–Neumann operator for $\mathbb{R}^3_-$ is given by

$$\partial_y \varphi(x,y)\Big|_{y=0} = |D_x|\xi(x) \ .$$

Therefore the equations of motion (7.24) for the linear water wave problem are given by

$$\partial_t \begin{pmatrix} \eta \\ \xi \end{pmatrix} = \begin{pmatrix} 0 & |D_x| \\ -g & 0 \end{pmatrix} \begin{pmatrix} \eta \\ \xi \end{pmatrix} + \begin{pmatrix} 0 \\ F \end{pmatrix} \ . \tag{7.25}$$

(ii) Give a solution to equations (7.24) using Fourier synthesis.

Suggestions: To put ourselves in the spirit of applied mathematics, we will denote the Fourier transform variable by $k = (k_1, k_2) \in \mathbb{R}^2$, known as the *wavenumber*. Take the Fourier transform of equations (7.25) in the horizontal spatial variables,

$$\partial_t \begin{pmatrix} \hat\eta \\ \hat\xi \end{pmatrix} = \begin{pmatrix} 0 & |k| \\ -g & 0 \end{pmatrix} \begin{pmatrix} \hat\eta \\ \hat\xi \end{pmatrix} + \begin{pmatrix} 0 \\ \hat F \end{pmatrix} \ . \tag{7.26}$$

Eigenvalues of the matrix of the RHS are $\pm i\omega(k)$, giving the dispersion relation, which in our case of the water wave equations in deep water is given by $\omega(k) = \sqrt{g|k|}$. This solution can be conveniently expressed in complex coordinates, writing

$$\hat z(t,k) = \frac{1}{\sqrt{2}} \left(\sqrt[4]{\frac{g}{|k|}}\,\hat\eta + i \sqrt[4]{\frac{|k|}{g}}\,\hat\xi \right) \ .$$

In these variables, the problem is reduced to solving the resulting complex scalar equation for $\hat{z}(t,k)$,

$$\partial_t \hat{z}(t,k) = i\omega(k)\hat{z}(t,k) + \frac{i}{\sqrt{2}}\sqrt[4]{\frac{|k|}{g}}\hat{F}(t,k) \; . \tag{7.27}$$

When $\hat{F}$ is nonzero, this can be solved using Duhamel's principle.

(iii) The perturbation of the free surface due to the presence of a ship is modeled by a pressure source F. A ship moving at constant speed U is taken to be the source of a uniformly moving Dirac δ-function, $p\delta(x - Ut)$. Show that after Fourier transform, the auxiliary force on the surface is expressed by

$$\hat{F}(t,k) = \frac{p}{2\pi} e^{i(k\cdot Ut)} \; .$$

Then show that the solution of (7.27) is given by

$$\hat{z}(t,k) = e^{i\omega(k)t}\hat{z}(0,k) + \frac{ip}{2\sqrt{2}\pi}\sqrt[4]{\frac{|k|}{g}}\Big(\frac{e^{ik\cdot Ut} - e^{i\omega(k)t}}{i(k\cdot U - \omega(k))}\Big) \; .$$

(iv) The inverse Fourier transform of this expression gives the solution in complex coordinates in the spatial variables. Transforming to coordinates in which the ship is stationary, $X = x + Ut$, show that the solution is in the form

$$\begin{aligned} z(t,X) = {} & \frac{1}{2\pi}\int e^{ik\cdot(X-Ut)} e^{i\omega(k)t}\hat{z}(0,k)\,dk \\ & + \frac{ip}{4\sqrt{2}\pi^2}\int \sqrt[4]{\frac{|k|}{g}}\Big(\frac{e^{ik\cdot X} - e^{i(k\cdot(X-Ut)+\omega(k)t)}}{i(k\cdot U - \omega(k))}\Big)\,dk \; . \end{aligned}$$

(v) Instead of solving the initial value problem from $t = 0$, suppose that motion started at some asymptotically large past time $-T$, from which we assume that all time-dependent transient waves are now negligible. In this situation the stationary wave pattern of the ship wake is given by

$$z(X) = \frac{ip}{4\sqrt{2}\pi^2}\int \sqrt[4]{\frac{|k|}{g}}\Big(\frac{e^{ik\cdot X}}{i(k\cdot U - \omega(k))}\Big)\,dk \; .$$

The integrand has a denominator that vanishes along two curves. Assume that $U = (U_1, 0)$, parametrize these curves in polar coordinates in the Fourier transform variables as $(k_1(\alpha), k_2(\alpha)) = (\kappa(\alpha)\cos(\alpha), \kappa(\alpha)\sin(\alpha))$ where

$$\kappa(\alpha) = \frac{g}{U_1^2}\frac{1}{\cos(\alpha)} \; ,$$

and use the calculus of residues to express this two-dimensional integral as a line integral.

(vi) The line integrals from part (v) take the form

$$z(X) = \int_{-\pi/2}^{+\pi/2} e^{i\Theta(\alpha,X)} A(\alpha)\, d\alpha \ ,$$

where the phase function is given by

$$\Theta(\alpha, X) = \kappa(\alpha)\cos(\alpha)X_1 + \kappa(\alpha)\sin(\alpha)X_2 \ .$$

In addition, the spatial variable can be represented in polar coordinates, $X = (X_1, X_2) = (r\cos(\beta), r\sin(\beta))$. For fixed X, show that the point of stationary phase occurs when

$$\tan(\alpha) + (1 + 2\tan^2(\alpha))\tan(\beta) = 0 \ .$$

Observe that this equation has two real roots when

$$|\tan(\beta)| \le \frac{1}{2\sqrt{2}} \ ,$$

which are distinct when the inequality is strict. These $X = (r\cos(\beta), r\sin(\beta))$ are in the region of the wake behind the ship. In the interior of the wake region, two distinct wavenumbers are dominant at each point X.

(vii) As mentioned in section 7.4, points at which the derivative of the phase degenerates have the potential to be more prominent features in the solution. This is indeed the case with the wake of a ship. Show that the second derivative of the phase function $\Theta(\alpha, X)$ with respect to α vanishes for

$$(1 + \tan^2(\alpha))(1 + 4\tan(\alpha)\tan(\beta)) = 0 \ .$$

Therefore both $\partial_\alpha\Theta(\alpha, X) = 0$ and $\partial_\alpha^2\Theta(\alpha, X) = 0$ when

$$|\tan(\beta)| = \frac{1}{2\sqrt{2}} \sim 19.47^\circ \ ,$$

that is to say, at the outer boundaries of the wake region described above.

Conclusion: The fact that people find remarkable is that this angle of the most prominent features of a ship wake is independent of the speed $|U|$ of the ship. In addition, the wave number $k(\alpha)$ associated with the waves produced at this angle is such that $\alpha = \arctan(1/\sqrt{2}) \sim 35.26^\circ$ is independent of the ship speed. What does change is the magnitude $|k(\alpha)|$, as well as other details of the wake. A continuation of this project is to discuss a similar analysis of the wake of a ship which is in motion in a body of water over a bottom of constant finite depth.

Chapter 8

Conservation laws and shocks

Shock waves in the interstellar medium in the Vela Constellation. Image credit and copyright: Angus Lau, Y. Van, SS Tong (Jade Scope Observatory).

8.1. First-order quasilinear equations

This section is devoted to the study of nonlinear first-order scalar equations and their solutions. We will focus on the case of one space and one time dimension. The first example of an equation of this form is Burger's equation

$$\begin{aligned} \partial_t u + u\,\partial_x u &= 0\ , \qquad x \in \mathbb{R}^1\ , \\ u(0,x) &= h(x)\ . \end{aligned} \tag{8.1}$$

Nonlinear equations where the derivatives of the unknown function enter only linearly, as they do in (8.1) above, are called *quasilinear.* This equation is reminiscent of the PDEs that were studied in Chapter 2 of the form

$$\partial_t u + c\,\partial_x u = 0\ ,$$

where c represents the speed of propagation. In quasilinear equations such as (8.1), the speed of propagation depends upon the unknown function $u(t,x)$, that is to say, on the solution itself.

Equations of the general form of (8.1) are called *conservation laws*, which is explained as follows. In physical settings, the function $u(t,x)$ represents the density of some quantity (such as a gas or a concentration of a component of a fluid). A second function $f = f(u)$ represents the *flux* of this quantity, which is assumed to depend upon the density u itself. Then the law of conservation of quantity dictates that over any interval $[a,b) \subseteq \mathbb{R}^1$,

$$\partial_t\Big(\int_a^b u(t,x)\,dx\Big) + f(u(t,b)) - f(u(t,a)) = 0\ . \tag{8.2}$$

Namely, the total quantity contained in the interval $[a,b)$ changes in time t only through flux out of the interval through the point $x = b$ and flux into the interval through the point $x = a$.

If $f(u(t,x))$ is differentiable at all $x \in [a,b)$, (8.2) can be rewritten as

$$\begin{aligned} 0 &= \partial_t\Big(\int_a^b u(t,x)\,dx\Big) + \Big(\int_a^b \partial_x f(u)\,dx\Big) \\ &= \Big(\int_a^b \partial_t u(t,x) + \partial_x f(u)\,dx\Big)\ . \end{aligned}$$

Of course if $u(t,\cdot) \in C^1(\mathbb{R}^1)$, then this holds for any interval $[a,b)$; therefore $u(t,x)$ satisfies the PDE

$$\partial_t u + \partial_x f(u) = 0\ ,$$

which is to say

$$\partial_t u + f'(u)\partial_x u = 0\ . \tag{8.3}$$

When the flux function is $f(u) = \frac{1}{2}u^2$, this equation is precisely Burger's equation (8.1).

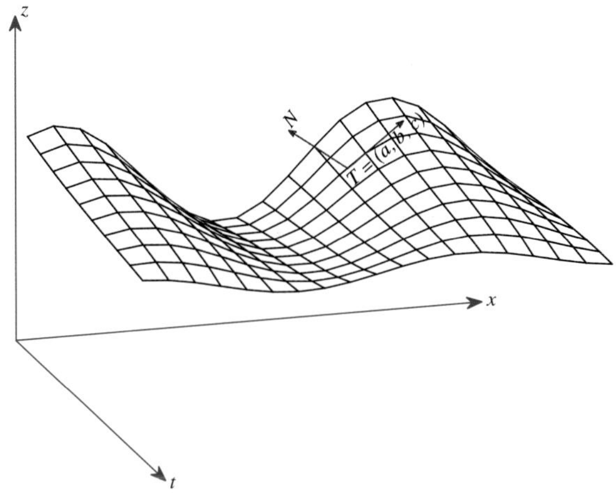

Figure 1. A solution to (8.5) as a graph $(t, x, u(t, x)) \subseteq (t, x, z)$.

General first-order quasilinear equations. When the initial data is smooth enough, $u(0, x) = h(x) \in C^1(\mathbb{R}^1)$, one solves these equations locally in time by first-order methods consisting of integrating along characteristics. The *characteristic equations* for Burger's equation are

$$\frac{dt}{ds} = 1 \ , \quad \frac{dx}{ds} = z \ , \quad \frac{dz}{ds} = 0 \ .$$

These characteristic ODEs can be solved explicitly:

$$s = t \ , \quad z(s) = z(0) \ , \quad x(s) = x(0) + sz \ . \tag{8.4}$$

This merits an explanation; we will give an interpretation in geometric terms of first-order quasilinear equations in general and their associated characteristic ODEs. The general quasilinear equation when $(t, x) \in \mathbb{R}^2$ is

$$a(t, x, u)\partial_t u + b(t, x, u)\partial_x u = c(t, x, u) \ . \tag{8.5}$$

A solution $u(t, x)$ is presented as a graph $(t, x, u(t, x))$ in $(t, x, z) \subseteq \mathbb{R}^3$, whose normal is

$$N = (-\partial_t u, -\partial_x u, 1) \ .$$

The coefficients of equation (8.5) are interpreted as a vector field $T = (a(t, x, z), b(t, x, z), c(t, x, z))$ in $(t, x, z) \in \mathbb{R}^3$ (the *characteristic vector field*), and equation (8.5) is the geometrical statement that

$$N \cdot T = 0 \ .$$

The situation is illustrated in Figure 1. This implies that the integral curves $(t(s), x(s), z(s))$ (the *characteristic curves*) of the characteristic vector field

$$\frac{dt}{ds} = a(t, x, z) \ , \quad \frac{dx}{ds} = b(t, x, z) \ , \quad \frac{dz}{ds} = c(t, x, z) \tag{8.6}$$

lie on the graph $z = u(t,x)$. A solution to (8.5) is given as the union of a parameter family of such characteristic curves. This family is locally specified by a condition that the characteristic curves pass through another given curve $\Gamma = (T(\beta), X(\beta), Z(\beta))$. In the above case of an initial value problem (8.1), this is the curve $\Gamma = (0, \beta, h(\beta))$ for $\beta \in \mathbb{R}^1$. To ensure that the solution surface is well defined, one requires that the tangent vectors to the curve Γ and the tangent vectors to the characteristic curves (8.6) at points on Γ are transversal, and that they locally span a tangent space which itself is a graph over the coordinate plane (t,x). This is the condition on the curve Γ that it be *noncharacteristic*. That is,

$$\det \begin{pmatrix} a & b \\ \partial_\beta T & \partial_\beta X \end{pmatrix} \neq 0 \ .$$

In the case (8.1) and for the initial value problem, this is to say that

$$\det \begin{pmatrix} 1 & z \\ 0 & 1 \end{pmatrix} \neq 0 \ ,$$

a condition which, in the case of Burger's equation, holds independently of the initial data $z = h(\beta)$.

Applying this construction to Burger's equation (8.4), we have derived the solution in the form

$$\begin{aligned} &u(t,x) = z(s,\beta) \ , \\ &\text{where} \quad t = s \ , \quad x(s,\beta) = \beta + sz \ , \quad z(s,\beta) = h(\beta) \ . \end{aligned} \tag{8.7}$$

Describing this solution process in words, the values of the solution $u(t,x)$ are given by $h(\beta)$ where $(\beta, h(\beta))$ is the point where the characteristic curve $(t(s,\beta), x(s,\beta), z(s,\beta))$ intersects the $\{t = 0\}$ coordinate plane. Let us assume that the initial data satisfies $h \in C^1(\mathbb{R}^1)$. To present the solution as a graph, we need to solve the above relations (8.7) for $u(t,x) = z(s,\beta)$. That is, one considers the Jacobian of the mapping $(s,\beta) \to (t,x)$, which is given by

$$\det \begin{pmatrix} t_s & x_s \\ t_\beta & x_\beta \end{pmatrix} = \det \begin{pmatrix} 1 & z(0,\beta) \\ 0 & 1 + s\partial_\beta z(s,\beta) \end{pmatrix} \ . \tag{8.8}$$

For small times $t = s$, the Jacobian will be invertible, but the fact is that for larger times t, the Jacobian determinant has the possibility to vanish. Indeed, this is the case whenever $1 + t\partial_\beta h(\beta) = 0$, in other words when

$$t = \frac{-1}{\partial_\beta h(\beta)} \ .$$

Therefore, for $h \in C^1(\mathbb{R}^1)$, $\partial_\beta h$ is bounded and there is an interval of time for which the above construction gives a C^1 solution to (8.1). However, at the first time t for which the Jacobian vanishes, the graph $z = u(t,x)$ will fold, developing a vertical tangent, and a discontinuity, or *shock*, will form.

Theorem 8.1. *If the initial data $h(x)$ for Burger's equation is decreasing at some point $x = \beta$, then the solution will become singular at a time $T^* > 0$, where*

$$T^* = \frac{-1}{\inf_\beta \partial_x h(\beta)} \ .$$

Proof. We will give two proofs of this result. The first follows from the Jacobian calculation that appears above. Let β_0 be a point which minimizes $\partial_x h(x)$. As $t \to T^* = -\big(\partial_x h(\beta_0)\big)^{-1}$, the Jacobian determinant (8.8) tends to zero as the tangent plane of the graph $z = u(t,x)$ becomes vertical. In the case where there is no point which achieves the minimum, nonetheless the Jacobian will tend to zero on a minimizing sequence (t_n, x_n) where $|x_n| \to +\infty$ and $t_n \to T^*$ as $n \to +\infty$, implying that the solution $u(t,x)$ has an unbounded C^1-norm in the limit.

A second proof of this theorem on the inevitable formation of shocks comes from an implicit relation satisfied by the solution. Namely, since $u = z(s,\beta) = z(0,\beta) = h(\beta)$, $t = s$, and $\beta = x - sz(s,\beta)$,

$$u(t,x) = h(\beta) = h(x - tu) \ .$$

Taking a derivative of this relation with respect to x,

$$\partial_x u(t,x) = h'(x - ut)(1 - t\partial_x u) \ ,$$

and solving for $\partial_x u$, we get

$$\partial_x u(t,x) = \frac{h'(\beta)}{1 + th'(\beta)} \ .$$

If $h'(\beta_0) < 0$ for some β_0, this expression becomes singular for $t = -(h'(\beta_0))^{-1}$. If β_0 is the infimum of $h'(\beta)$, then the first time $t > 0$ at which this expression becomes singular is precisely T^*. □

The outcome of these considerations is that we are forced to take into account a sense of a solution of the differential equations which is not necessarily even continuous.

Definition 8.2. The function $u(t,x)$ is a *weak solution* of Burger's equation (8.1) if it satisfies the original conservation law (8.2) for all intervals $[a,b) \subseteq \mathbb{R}^1$.

At points (t,x) at which $u \in C^1$, the notion of weak solution coincides with our usual sense of interpretation of the PDE. However, suppose that

$u(t,x)$ were discontinuous across a C^1 curve $x = \gamma(t)$. Then on every interval $[a,b)$ which contains a point on the curve, a weak solution must satisfy

$$\begin{aligned}
0 &= \partial_t\Big(\int_a^{\gamma(t)} u\,dx + \int_{\gamma(t)}^b u\,dx\Big) + f(u(t,b)) - f(u(t,a)) \\
&= \int_a^{\gamma(t)} \partial_t u\,dx + \dot\gamma(t)u_-(t,\gamma(t)) + \int_{\gamma(t)}^b \partial_t u\,dx - \dot\gamma(t)u_+(t,\gamma(t)) \\
&\qquad + f(u(t,b)) - f(u(t,a)) \\
&= \int_a^{\gamma(t)} \partial_t u\,dx + f(u_-(t,\gamma(t))) - f(u(t,a)) \\
&\qquad + \int_{\gamma(t)}^b \partial_t u\,dx + f(u(t,b)) - f(u_+(t,\gamma(t))) \\
&\quad + \dot\gamma(t)[u_-(t,\gamma(t)) - u_+(t,\gamma(t))] - [f(u_-(t,\gamma(t))) - f(u_+(t,\gamma(t)))]\ .
\end{aligned}$$

We have used the notation of the limit from the left $\lim_{x\to\gamma(t)_-} u(t,x) = u_-(t,x)$ and similarly for the right limit. Using the definition of a solution in the two regions of regularity $[a,\gamma(t))$ and $(\gamma(t),b]$, we obtain

$$0 = \dot\gamma(t)[u_-(t,\gamma(t)) - u_+(t,\gamma(t))] - [f(u_-(t,\gamma(t))) - f(u_+(t,\gamma(t)))]\ . \tag{8.9}$$

This is the *jump condition*, matching the speed of propagation of the discontinuity with the magnitude of the jump, that is to be satisfied in order to be a weak solution at a discontinuity.

Theorem 8.3. *A piecewise $C^1(\mathbb{R}^2_{tx})$ solution $u(t,x)$ of (8.3) which is discontinuous across a curve $x = \gamma(t) \in C(\mathbb{R}^1)$ is a weak solution if:*

(1) *it satisfies the PDE at a C^1 point, and*

(2) *across discontinuities it satisfies the jump condition (8.9), namely*

$$\dot\gamma[u_- - u_+] = [f(u_-) - f(u_+)]\ . \tag{8.10}$$

One could ask about the variety of scalar conservation laws for different flux functions $f(u)$. Consider the general case as in (8.3),

$$\partial_t u + f'(u)\partial_x u = 0\ .$$

Suppose that $u(t,x) \in C^1$ in some neighborhood $\Omega \in \mathbb{R}^2_{tx}$, and assume that $f(u)$ is strictly convex, which is to say that $f'(u)$ is a strictly increasing function in u. Then $g(v) = (f')^{(-1)}(v)$ is well defined, and the transformation $u = g(v)$ is such that in Ω, the function $v(t,x)$ satisfies

$$\partial_t u = g'(v)\partial_t v\ , \quad f'(u)\partial_x u = f'(g(v))g'(v)\partial_x v\ .$$

Since $f'(g(v)) = v$, $v(t,x)$ satisfies

$$\partial_t v + v\partial_x v = 0\ ,$$

which is of course Burger's equation. That is, for hyperbolic conservation laws with a convex flux function, for smooth solutions, all such equations are identical up to a change of dependent variable. This is not necessarily the case for weak solutions, however; Exercise 8.4 discusses the effect of changes of variables on the jump condition.

8.2. The Riemann problem

Under dilations $t' = \alpha t$, $x' = \alpha x$, the equation

$$\partial_t u + c(u)\partial_x u = 0$$

remains invariant. The class of solutions that are also invariant under dilations is particularly simple, and it plays an important role in the theory and applications of such conservation laws. In particular, initial data which is dilation invariant is given by

$$u(0,x) = h(x) \ , \quad \text{where } h(x) = h_- \text{ for } \ x < 0 \ , \quad h(x) = h_+ \text{ for } \ x \geq 0 \ ,$$

and where $h_\pm$ are two constants. This is called the *Riemann problem.* For Burger's equation

$$\partial_t u + u\partial_x u = 0 \ , \tag{8.11}$$

solutions are simple and explicit.

Case (1). Given $h_- = 0$ and $h_+ = 1$, for $t > 0$ a solution is

$$\begin{aligned} u(t,x) &= 0 \quad \text{for } \ x \leq 0 \ , \\ u(t,x) &= \frac{x}{t} \quad \text{for } \ 0 \leq x \leq t \ , \\ u(t,x) &= 1 \quad \text{for } \ x \geq 0 \ . \end{aligned}$$

Such solutions are *rarefaction waves*, as they propagate a decrease in the density of the quantity u. A solution with a rarefaction wave is illustrated by its characteristic curves in Figure 2. Notice that the solution $u(t,x)$ is not $C^1(\mathbb{R}^2)$ for $t > 0$ but $u \in Lip(\mathbb{R}^2)$.

Case (2). Given $h_- = 1$ and $h_+ = 0$, for $t > 0$ a solution is necessarily discontinuous; indeed,

$$\begin{aligned} u(t,x) &= 1 \quad \text{for } \ x < \frac{t}{2} \ , \\ u(t,x) &= 0 \quad \text{for } \ x \geq \frac{t}{2} \ . \end{aligned}$$

A solution with a shock wave is illustrated in Figure 3. The line of discontinuity given by $x = \frac{t}{2}$ is a *shock* curve. Its speed is determined by the jump

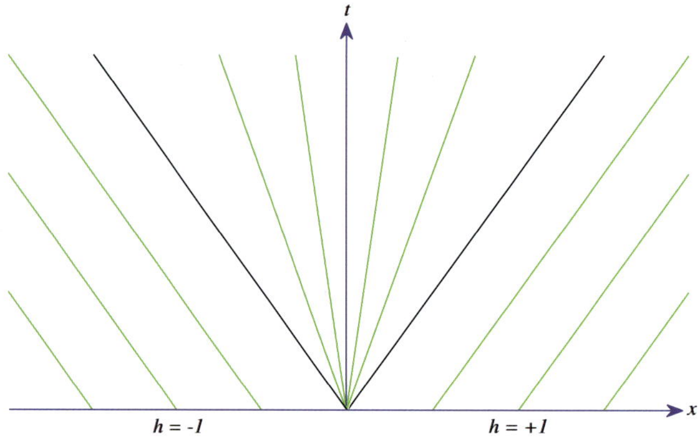

Figure 2. A rarefaction wave for Burger's equation (8.11).

condition (8.10); indeed, we check this:

$$u_- = 1 \ , \quad u_+ = 0 \ ,$$

$$u_- - u_+ = 1 \ , \quad f(u_-) - f(u_+) = \frac{1}{2}u_-^2 - \frac{1}{2}u_+^2 = \frac{1}{2} \ ; \quad \text{thus } \dot\gamma(t) = \frac{1}{2} \ .$$

Case (3). There is another case to consider. Namely, starting with the initial data of Case (1), $h_- = 0$ and $h_+ = 1$, there is a discontinuous solution that satisfies all of the conditions that we have so far imposed on solutions. Namely,

$$u(t,x) = 0 \quad \text{for} \quad x < \frac{t}{2} \ ,$$
$$u(t,x) = 1 \quad \text{for} \quad x \geq \frac{t}{2} \ .$$

Checking the jump condition (8.10), we have

$$u_- = 0 \ , \quad u_+ = 1 \ ,$$

$$u_- - u_+ = -1 \ , \quad f(u_-) - f(u_+) = \frac{1}{2}u_-^2 - \frac{1}{2}u_+^2 = -\frac{1}{2} \ ; \quad \text{thus } \dot\gamma(t) = \frac{1}{2} \ .$$

This is a primary example of nonuniqueness of weak solutions. In relaxing the definition of a solution in order to accommodate discontinuities, we have increased the class of allowable solutions, and in doing so have lost the property of uniqueness of solutions for given initial data. In order to regain it, one must make a selection of the appropriate solution by imposing a further principle, in addition to the definition of a weak solution through the law of conservation.

Definition 8.4 (Entropy condition)**.** If $u(t,x)$ is a weak solution of (8.3) which is piecewise C^1, and $x = \gamma(t)$ is a curve along which it has a jump

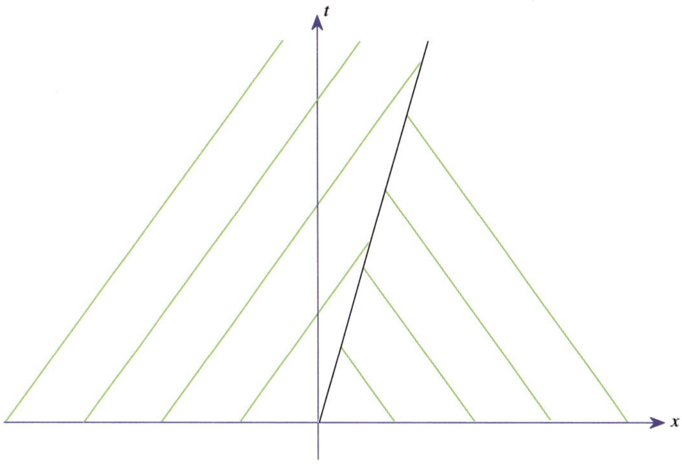

Figure 3. A shock wave for Burger's equation (8.11).

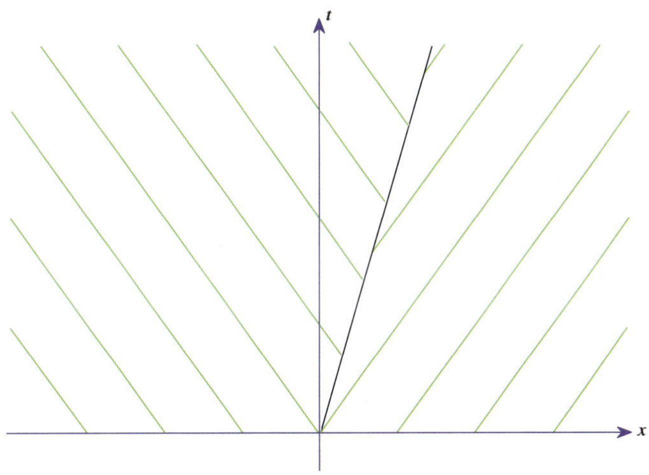

Figure 4. A shock wave for Burger's equation (8.11) that does not satisfy the entropy condition.

discontinuity, we define $u(t, x)$ to be an *entropy solution* if

$$f'(u_-) := c(u_-) > \dot{\gamma} \quad \text{and} \quad f'(u_+) := c(u_+) < \dot{\gamma} \ . \tag{8.12}$$

A solution that does not satisfy the entropy condition is illustrated in Figure 4. This condition brings back uniqueness to the initial value problem, as we will see in the following section. It is important to understand that this is an additional condition that has been added to the evolution problem for (8.3), above and beyond the concept of a weak solution to a conservation law. In particular, this choice of acceptable solution has introduced a distinction between future and past, a so-called *arrow of time*. It means that as time increases towards the future, characteristic curves which neighbor a shock curve must run into the shock. The solution of the Riemann problem for Burger's equation in Case (2) satisfies the entropy condition, while the solution of the Riemann in Case (3) does not satisfy it, and characteristics can and do originate on the shock curve and propagate away from it. In terms of entropy, for a shock in a solution that satisfies the entropy condition, information carried along characteristics of a solution will pass into the shock and disappear. However, in Case (3), information could be generated in the shock curve itself and propagate into the future; indeed, another possible weak solution for the initial data given in Case (3) is a modification of the shock curve $\gamma(t) = t/2$, for which at each second after $t > 0$ (and for some small value of β), the shock curve satisfies $\dot{\gamma} = 1/2 \pm \beta$. Choosing the sign of $\pm\beta$, each subsequent second gives a coding in binary, and in principle we may encode all of the Oxford English Dictionary in a weak solution of Burger's equation, which itself is one of an infinite number of nonunique solutions. Thus it makes sense to impose the entropy condition as a selection principle for weak solutions.

8.3. Lax–Olenik solutions

So far, for scalar hyperbolic conservation laws we have constructed solutions locally in time using the method of characteristics, and for special Riemann initial data we have constructed quite explicit solutions that exist globally in time and that exhibit shocks and rarefaction waves. There is a beautiful extension of these considerations that gives global weak solutions for general $L^1(\mathbb{R}^1)$ initial data, solutions which satisfy the entropy condition of Definition 8.4. Furthermore, uniqueness holds for this class of solutions. A nice reference for this material is by Lax [**14**].

Consider a general scalar conservation law for $f \in C^2$

$$\partial_t u + \partial_x f(u) = 0 ,$$

which we rewrite by the chain rule as

$$\partial_t u + a(u)\partial_x u = 0 , \tag{8.13}$$

where evidently $a(u) = \partial_u f(u)$. Without loss of generality, we may assume that $f(0) = 0$. We take up the study of the initial value problem for initial data $h(x) = u(0, x) \in L^1(\mathbb{R}^1)$. The principal hypothesis is that the flux function $f(u)$ is strictly convex; this assures that the coefficient $a(u)$ is monotone increasing in u. This hypothesis is satisfied for example in the case of Burger's equation, where $f(u) = \frac{1}{2}u^2$.

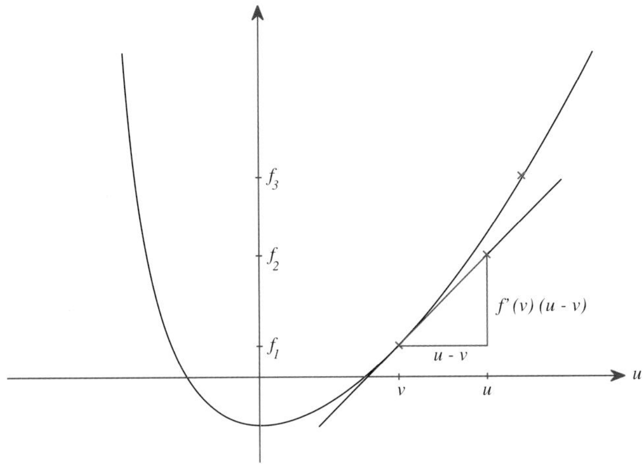

Figure 5. Property of convexity, as described in inequality (8.14).

Convexity and convex duality. The properties of $f(u)$ and $a(u)$ are used in several ways. Firstly, the following inequality holds for arbitrary u and v:

$$f(v) + a(v)(u - v) \leq f(u) \ . \tag{8.14}$$

This inequality is a consequence of convexity, as illustrated in Figure 5. Because of strict convexity, equality holds only for $u = v$. Being convex, $f(v)$ has a *convex dual* defined by

$$g(z) := a(v)v - f(v) \ , \quad \text{where} \quad z = a(v) \ .$$

The dual has the alternative definition

$$g(z) = \min_v \big(zv - f(v)\big) \ ,$$

where the minimum is taken on when $\partial_v\big(zv - f(v)\big) = 0$, namely for $z - \partial_u f(v) = 0$. Figure 6 illustrates this relationship. Since $a(v)$ is increasing in v, we may define its inverse function, setting $v = b(z)$ whenever $a(v) = z$. Incidentally, when $v = b(z)$, z is the minimizer of the function

$$f(v) = \min_z \big(zv - g(v)\big) \ ,$$

namely $v = \partial_z g(z) = b(z)$. It follows from a short calculation that $g(z)$ is convex.

This is similar in the setting of a Lagrangian $L(\dot{u}, u)$ to the Legendre transform, which is globally defined when L is convex in the variables $\dot{u}$. Examples of convex functions and their convex duals are given by

$$f(v) = \frac{A}{2}v^2 \ , \quad g(z) = \frac{1}{2A}z^2 \ ,$$

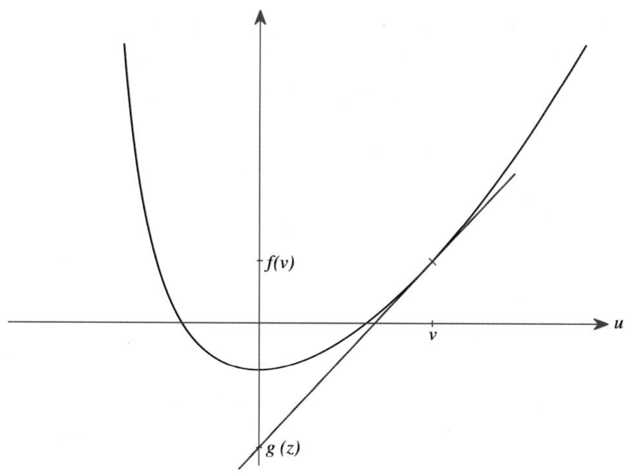

Figure 6. Convex duality.

and

$$f(v) = \frac{1}{p} v^p \ , \quad g(z) = \frac{1}{p'} z^{p'} \ , \quad \text{where} \quad \frac{1}{p} + \frac{1}{p'} = 1 \ ,$$

all of which are straightforward to check.

Hamilton–Jacobi equations. It is useful to change the problem to the setting of a Hamilton–Jacobi equation, namely to study

$$U(t, x) = \int_{-\infty}^{x} u(t, x')\, dx' \ ,$$

so that $\partial_x U = u$. Then U satisfies

$$\partial_t U + f(\partial_x U) = 0 \ . \tag{8.15}$$

Initial data is given by

$$H(x) = \int_{-\infty}^{x} h(x')\, dx' \ ,$$

which is Lipschitz continuous since we have asked that $h \in L^1$. Furthermore, for almost all x_1, we have[1]

$$H(x_2) - H(x_1) = h(x_1)(x_2 - x_1) + o(|x_2 - x_1|) \ .$$

Whenever $u(t, x) \in C^1(\mathbb{R}^1)$, $U(t, x)$ is a classical solution to (8.15). By convexity, given a solution U of (8.15), for any v, (8.14) implies that

$$\partial_t U + \partial_u f(v)\partial_x U = \partial_u f(v)\partial_x U - f(\partial_x U) \le \partial_u f(v) v - f(v) \ . \tag{8.16}$$

Because of strict convexity, equality holds only for $v = \partial_x U$.

[1]We use big $\mathcal{O}(\cdot)$ and little $o(\cdot)$ notation.

For purposes of constructing a solution to equation (8.15), given a space-time point (t, x) with $t > 0$, consider the line in space-time through it with slope $a(v) = \partial_u f(v)$; the intersection of this line with the x-axis defines y such that

$$\frac{x-y}{t} = a(v) = z \ .$$

The LHS of inequality (8.16) has the interpretation as the directional derivative of the function $U(t,x)$ along this line, while the RHS depends only on the parameter v. Integrating (8.16) along this line gives that

$$U(t,x) \le U(0,y) + t\big(a(v)v - f(v)\big) = H(y) + t\big(a(v)v - f(v)\big) \ . \tag{8.17}$$

Interpret this inquality in terms of convex duality; take $z = a(v)$ and $g(z) = a(v)v - f(v)$ as above. That is, $b(z) = v$, the inverse function, so that when

$$z = \frac{x-y}{t} = a(v) \ , \quad \text{then} \quad b(z) = b\big(\frac{x-y}{t}\big) = v \ .$$

This gives a reinterpretation of inequality (8.17), articulated in the following proposition.

Proposition 8.5. *If $u(t,x)$ is a solution of* (8.13) *which is Lipschitz at $x \in \mathbb{R}^1$ at time $t > 0$, then*

$$u(t,x) = b\big(\frac{x-y}{t}\big) = h(y(t,x)) \ , \tag{8.18}$$

where $y = y(t,x)$ minimizes the relation involving the initial data $H(y)$:

$$H(y) + tg\big(\frac{x-y}{t}\big) := G(x,y;t) \ . \tag{8.19}$$

The ray $(x - y(t,x))/t$ is the backwards characteristic between (t,x) and $(0, y(t,x))$ over the time interval $[0,t]$.

Lax–Olenik weak solutions. Now suppose that $u(t,x)$ is a weak solution with possible jump discontinuities. Then nonetheless

$$U(t,x) = \int_{-\infty}^{x} u(t,x')\, dx' \ ,$$

which is Lipschitz (as long as $u(\cdot, x) \in L^1$), and the inequality still holds that

$$\partial_t U + \partial_u f(v) \partial_x U \le \partial_u f(v) v - f(v) \ .$$

If $u(t,x)$ is a weak solution that in addition satisfies the entropy condition (8.12) at each discontinuity,

$$u_+ \le \frac{d\gamma}{dt} \le u_- \ ,$$

then every point $(t,x) \in \mathbb{R}_t^+ \times \mathbb{R}_x$ is connected to the initial data at $t = 0$ by at least one ray $(x-y)/t = a(v)$ since backward characteristics do not intersect shock curves. For each minimizing value of y, equality holds in (8.16).

The elegant aspect of the theory is that the converse holds. Namely, the variational formula given by (8.18) and (8.19) defines a weak solution to (8.13) that in addition satisfies the entropy condition at every discontinuity.

Theorem 8.6. *Formulae* (8.18) *and* (8.19) *define a possibly discontinuous function* $u(t,x)$ *for arbitrary initial data* $h(x) \in L^1(\mathbb{R}^1)$, *which is a weak solution with at most countably many jump discontinuities, and which satisfies the entropy condition at such discontinuities.*

The function $u(t,x)$ defined by (8.18) and (8.19) is called the *Lax–Olenik* solution.

Proof. First of all, because $h(x) \in L^1$, $H(x)$ is bounded, and the convex dual $g(z) = a(v)v - f(v)$ for $v = b(z)$ taken over the range of $H(y)$ has its minimum at some $z = c$ where $\partial_z g(z) = 0$. Therefore the function $G(x,y;t)$ defined in (8.19) is minimized and takes on its minimum at some finite y. This minimizer is possibly nonunique. If y is the unique minimizer of $G(x,y;t)$, then $\partial_y G(x,y;t) = 0$, at least for *a.e.* y at which H' exists, which implies that

$$0 = H'(y) - g'\left(\frac{x-y}{t}\right) = h(y) - b\left(\frac{x-y}{t}\right) .$$

In fact this $y = y(t,x)$ is the minimizer for all points on the ray between $(0,y)$ and (t,x); indeed, for $0 < t_1 < t$ and for x_1 such that

$$\frac{x_1 - y}{t_1} = \frac{x-y}{t} ,$$

$\partial_y G(x_1, y; t_1) = 0$ so that $h(y) = b(\frac{x_1-y}{t_1})$ as well. Thus the PDE (8.13) is satisfied along this ray, and $u(t,x) = h(y) = u(t_1, x_1)$.

However, it could be that for given (t,x), $G(x,y;t)$ has more than one minimizer $y(t,x)$; this situation is illustrated in Figure 8. In this case, the following property of $y(t,x)$ is useful.

Lemma 8.7. *Fix* $t > 0$ *and denote* $y = y(t,x)$ *the minimizers of* $G(x,y;t)$. *Then* $y(t,x)$ *is a nondecreasing function of* x.

Proof of Lemma 8.7. The result will follow from a fact that we shall prove geometrically. Take $x_2 > x_1$, then $G(x_2, y_1) < G(x_2, y)$ for all $y < y_1$. This inequality depends upon the convexity of the function g:

$$g\left(\frac{x_2 - y_1}{t}\right) + g\left(\frac{x_1 - y}{t}\right) < g\left(\frac{x_1 - y_1}{t}\right) + g\left(\frac{x_2 - y_1}{t}\right) . \tag{8.20}$$

The intervals $[(x_2 - y_1)/t, (x_1 - y)/t]$ and $[(x_1 - y_1)/t, (x_2 - y)/t]$ share the same midpoint, and by convexity, the sketch in Figure 7 proves the inequality.

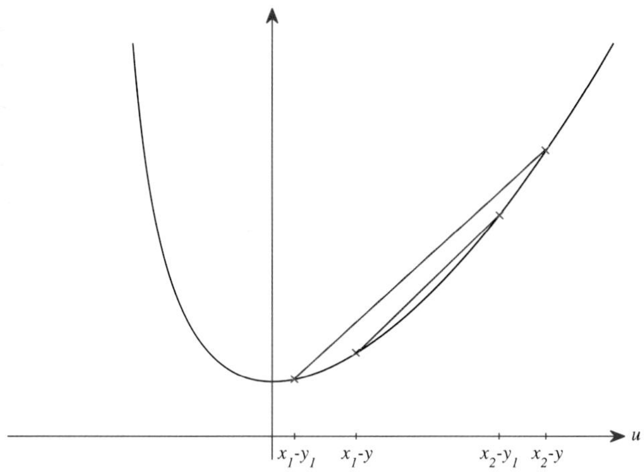

Figure 7. Geometric proof of inequality (8.20).

Recall that $G(x, y; t) = H(y) + tg((x - y)/t)$; multiply inequality (8.20) by t and respectively add $G(x_1, y_1; t) \leq G(x_1, y; t)$ to each side, obtaining

$$t\Big(g\big(\frac{x_2 - y_1}{t}\big) + g\big(\frac{x_1 - y}{t}\big)\Big) + G(x_1, y_1; t)$$
$$\leq t\Big(g\big(\frac{x_1 - y_1}{t}\big) + g\big(\frac{x_2 - y}{t}\big)\Big) + G(x_1, y; t) \ .$$

Using definition (8.19) for $G(x, y; t)$, this is the statement that

$$tg\big(\frac{x_2 - y_1}{t}\big) + H(y_1) \leq tg\big(\frac{x_2 - y}{t}\big) + H(y) \ ,$$

which is to say that for $y < y_1$,

$$G(x_2, y_1; t) \leq G(x_2, y; t) \ .$$

Thus if y_1 is a minimizer for $x = x_1$, then any minimizer for $x_2 > x_1$ must satisfy $y_2 > y_1$. □

The first conclusion we draw from this monotonicity is that for fixed $t > 0$, the mapping $x \mapsto y(t, x)$ is increasing in x with only countably many jump discontinuities. Suppose that $G(x, y; t)$ has more than one minimizer, and let y^- be the least minimizer of $G(x, y; t)$ and $y^+ > y^-$ the greatest. Because of the property of monotonicity, for fixed $t > 0$ if $x_1 < x$, then any y_1 that is a minimizer of $G(x_1, y; t)$ must satisfy $y_1 < y^-$, while if $x_1 > x$, then $y_1 > y^+$. Therefore the interval $[y^-, y^+]$ is a jump discontinuity of $y(t, x)$, and it follows that there can only be countably many points x for which the minimizer $y = y(t, x)$ is not unique.

Now consider $t_1 > t$; for all $x_1 \in \mathbb{R}^1$, a minimizer $y(t_1, x_1)$ either satisfies $y(t_1, x_1) < y^-(t, x)$ or else $y^+(t, x) < y(t_1, x_1)$. This is because for any $y^-(t, x) < y < y^+(t, x)$, the ray from (t_1, x_1) to $(0, y)$ will intersect one

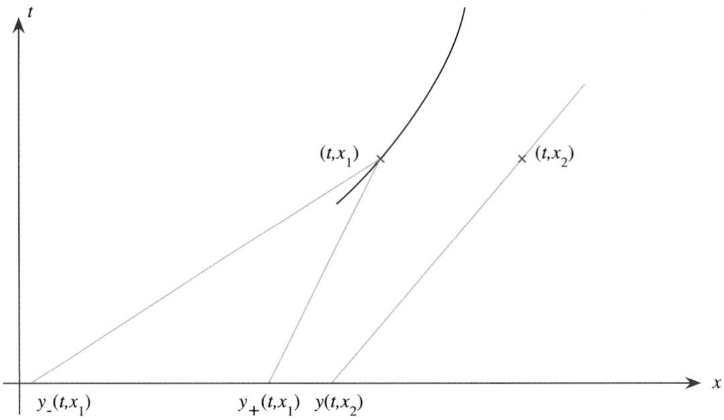

Figure 8. The characteristics in cases of nonunique (t, x_1) and unique (t, x_2) minimizers of the Lax–Olenik variational principle.

of the rays (t, x) to $(0, y^{\pm}(t, x))$ and therefore cannot have the minimizing slope of the latter. Hence when $t_1 > t$, for each interval $[y^-(t, x), y^+(t, x)]$ of jump discontinuity of $y(t, \cdot)$, there is a point x_1 with multiple minimizers of $G(x_1, y; t_1)$ and a corresponding interval $[y^-(t_1, x_1), y^+(t_1, x_1)]$ representing a jump in $y(t_1, x)$. In particular, if $\{x^k\}$ and $\{x_1^j\}$ are the points of discontinuity of $y(t, x)$ and $y(t_1, x)$ at times t and $t_1 > t$, respectively, then

$$\bigcup_k [y^-(t, x^k), y^+(t, x^k)] \subseteq \bigcup_j [y^-(t_1, x_1^j), y^+(t_1, x_1^j)] \ .$$

The union of these points with nonunique minimizers form a set of *shock curves*, which by the above property may merge and come into existence, but may not split as t increases. The presence of such space-time points for which $G(x, y; t)$ has nonunique minimizers implies that a solution given through the Lax–Olenik variational principle is not a classical solution. We now show that it is a weak solution in the sense that it satisfies the jump condition (8.9) at points on shock curves.

Consider $t_1 > t$ and points (t, x) and (t_1, x_1) on a shock curve, where each has nonunique minimizers $y^{\pm}$ and $y_1^{\pm}$, respectively, and such that

$$[y^-(t, x), y^+(t, x)] \subseteq [y_1^-(t_1, x_1), y_1^+(t_1, x_1)] \ .$$

Since they are nonunique minimizers, from the Lax–Olenik variational principle we know that

$$G(x, y^-; t) = G(x, y^+; t) \ , \quad G(x_1, y_1^-; t_1) = G(x_1, y_1^+; t_1) \ , \tag{8.21}$$

where of course $G(x, y; t) = H(y) + tg(\frac{x-y}{t})$. Interpreting (8.21) in terms of $f(v)$, and denoting $v^{\pm} = b(z^{\pm}) = b(\frac{x-y^{\pm}}{t})$, $v_1^{\pm} = b(z_1^{\pm}) = b(\frac{x_1-y_1^{\pm}}{t_1})$, equalities (8.21) become

$$t[f(v^+) - f(v^-)] - [v^+(x - y^+) - v^-(x - y^-)] = H(y^+) - H(y^-) \tag{8.22}$$

and

$$t_1[f(v_1^+) - f(v_1^-)] - [v_1^+(x_1 - y_1^+) - v_1^-(x_1 - y_1^-)] = H(y_1^+) - H(y_1^-) \,. \tag{8.23}$$

Taking the difference of (8.23) and (8.22) and using that

$$H(y_1^\pm) - H(y^\pm) = h(y^\pm)(y_1^\pm - y^\pm) + o(y_1^\pm - y^\pm) \,,$$

we find

$$\begin{aligned}(t_1 - t)[f(v^+) - f(v^-)] - (x_1 - x)[v^+ - v^-] \\ = -t_1[f(v_1^+) - f(v^+) - f(v_1^-) + f(v^-)] \\ + [(v_1^+ - v^+)(x_1 - y^+) - (v_1^- - v^-)(x_1 - y^-)] + o(y_1^\pm - y^\pm) \,.\end{aligned} \tag{8.24}$$

Now

$$f(v_1^\pm) - f(v^\pm) = a(v^\pm)(v_1^\pm - v^\pm) + o(v_1^\pm - v^\pm) \,;$$

therefore the RHS of (8.24) is given by

$$\begin{aligned} &- t_1[a(v^+)(v_1^+ - v^+) - a(v^-)(v_1^- - v^-)] \\ &\quad - [(v_1^+ - v^+)(x_1 - y^+) - (v_1^- - v^-)(x_1 - y^-)] \\ &\quad + o(v_1^\pm - v^\pm) + o(y_1^\pm - y^\pm) \\ &= -t_1[a(v^+)(v_1^+ - v^+) - a(v^-)(v_1^- - v^-) \\ &\quad - (v_1^+ - v^+)(\frac{x - y^+}{t}) + (v_1^- - v^-)(\frac{x - y^-}{t})] \\ &\quad + o(v_1^\pm - v^\pm) + o(y_1^\pm - y^\pm) + o(x_1 - x) + o(t_1 - t) \,. \end{aligned}$$

Observing that $a(v^\pm) = \frac{x - y^\pm}{t}$ and that $o(v_1^\pm - v^\pm)$ and $o(y_1^\pm - y^\pm)$ are $o(x_1 - x) + o(t_1 - t)$, we have shown that the LHS is $o(x_1 - x) + o(t_1 - t)$. This is tantamount to the jump condition (8.9), and hence the function $u(t, x)$ defined through the Lax–Olenik variational principle ((8.18) and (8.19)) is indeed a weak solution of (8.13).

It remains to verify that the solution given by (8.18) and (8.19) in Proposition 8.5 satisfies the entropy condition for the conservation law (8.13). Use again that $f \in C^2$ is convex so $b(z) \in C^1$ is increasing, and that $y = y(t, x)$ is increasing in x. Therefore for $x_1 < x_2$,

$$\begin{aligned} u(t, x_1) = b(\frac{x_1 - y_1}{t}) &\geq b(\frac{x_1 - y_2}{t}) \\ &\geq b(\frac{x_2 - y_2}{t}) - C_b\frac{x_2 - x_1}{t} \\ &= u(t, x_2) - C_b\frac{x_2 - x_1}{t} \,, \end{aligned} \tag{8.25}$$

where one uses in the second line that $b(z)$ has Lipschitz constant C_b:

$$b(z_2) - b(z_1) \leq C_b(z_2 - z_1) \quad \text{for} \quad z_1 \leq z_2 \ .$$

Therefore approaching a point x of discontinuity of $u(t,x) = b((x-y)/t)$ from the left by x_1 and from the right by x_2, our solution satisfies

$$u^-(t,x) \geq u^+(t,x) \ ,$$

which is the entropy condition for a jump discontinuity. □

Uniqueness. The solution satisfying the entropy condition constructed by the Lax–Olenik variational principle is unique, as we show with an argument involving contractions in L^1. Consider two initial conditions $h_1(x), h_2(x) \in L^1$ and their resulting Lax–Olenik solutions $u_1(t,x)$ and $u_2(t,x)$. We will show that the difference in L^1-norm of the two solutions is a decreasing function of time. The L^1-norm of their difference is

$$\begin{aligned} \|u_1(t,\cdot) - u_2(t,\cdot)\|_{L^1} &= \int_{-\infty}^{+\infty} |u_1(t,x) - u_2(t,x)| \, dx \\ &= \sum_n (-1)^n \int_{y_n(t)}^{y_{n+1}(t)} (u_1(t,x) - u_2(t,x)) \, dx \ , \end{aligned}$$

where $y_n(t)$ are the points of sign changes of the difference $u_1(t,x) - u_2(t,x)$, with $\mathrm{sgn}(u_1(t,x) - u_2(t,x)) = (-1)^n$ over the interval $(y_n(t), y_{n+1}(t))$. We will analyze the behavior of the individual integrals of this sum.

Case (1). Suppose that $y_n(t)$ is a point of continuity for both $u_1(t,x)$ and $u_2(t,x)$. In this case, both these solutions are constant along characteristics, which in turn are linear. Therefore $y_n(t)$ is linear in t, and $u_1(t, y_n(t)) - u_2(t, y_n(t)) = 0$.

Case (2). When $y_n(t)$ is a point of discontinuity of either $u_1(t,x)$ or $u_2(t,x)$, or both, $y_n(t)$ is a shock curve.

Using the nature of conservation laws and their weak solutions, the time derivative of the difference satisfies

$$\begin{aligned} \frac{d}{dt}\|u_1(t,x) - u_2(t,x)\|_{L^1} &= \sum_n (-1)^n \frac{d}{dt} \int_{y_n(t)}^{y_{n+1}(t)} (u_1(t,x) - u_2(t,x)) \, dx \\ &+ \sum_n (-1)^n \int_{y_n(t)}^{y_{n+1}(t)} \partial_t (u_1(t,x) - u_2(t,x)) \, dx \\ &+ \sum_n (-1)^n \Big[(u_1(t, y_{n+1}(t)) - u_2(t, y_{n+1}(t))) \frac{dy_{n+1}}{dt} \\ &\qquad - u_1(t, y_n(t)) - u_2(t, y_n(t)) \frac{dy_n(t)}{dt} \Big] \ . \end{aligned}$$

Each term of the sum after integration takes the form

$$\begin{aligned}(-1)^n \frac{d}{dt}\int_{y_n(t)}^{y_{n+1}(t)} (u_1(t,x) - u_2(t,x))\,dx \\ = (-1)^n\Big(\big(f(u_1(t,\cdot))-f(u_2(t,\cdot))\big)\Big|_{y=y_n(t)}^{y_{n+1}(t)}\Big)+\big(u_1(t,y)-u_2(t,y)\big)\frac{dy}{dt}\Big|_{y=y_n(t)}^{y_{n+1}(t)} .\end{aligned}$$

For endpoints $y_n(t)$ in Case (1), $u_1(t, y_n) = u_2(t, y_n)$ and there is no contribution to the sum. For endpoints in Case (2), there are subcases of which we will do one example. Suppose that it is $u_1(x,t)$ that has an entropy condition satisfying jump at $y = y_{n+1}(t)$, at which $u_1^+ \le u_2 \le u_1^-$, while u_2 is continuous. Then over the left-hand interval (y_n, y_{n+1}), we know that $u_1(t,x) - u_2(t,x) > 0$, so n is even. The jump condition at $y_{n+1}(t)$ is that

$$\frac{dy}{dt} = \frac{f(u_1^-) - f(u_1^+)}{u_1^- - u_1^+} ,$$

and by the convexity of $f(u)$, we have

$$\begin{aligned} f(u_2) - f(u_1^-) + (u_1^- - u_2)\Big(\frac{f(u_1^-) - f(u_1^+)}{u_1^- - u_1^+}\Big) \\ \le f(u_2) - \Big(\frac{u_2 - u_1^+}{u_1^- - u_1^+} f(u_1^-) + \frac{u_1^- - u_2}{u_1^- - u_1^+} f(u_1^+)\Big) \le 0 .\end{aligned}$$

Hence this boundary term gives a negative contribution:

$$(-1)^n (u_1 - u_2)(y)\frac{dy}{dt}\Big|_{y=y_{n+1}(t)} \le 0 .$$

Inspecting the other cases, they also only give negative contributions. Hence the L^1-norm of the difference is decreasing. In the case that the initial data satisfies $h_1 = h_2$, this implies the uniqueness of entropy condition-satisfying solutions and thus uniqueness for the class of Lax–Olenik solutions.

Exercises: Chapter 8

Exercise 8.1. Suppose that $f(x,u) = b(x)u$ is the flux function for a linear scalar conservation law, and consider a solution $u(t,x) \in C^1(\mathbb{R}^2)$:

$$\partial_t u + \partial_x (b(x)u) = 0 .$$

Suppose that $x = X_1(t)$ and $x = X_2(t)$ are two characteristic curves such that $X_1(t) < X_2(t)$. Show that for all $t \in \mathbb{R}^1$,

$$\int_{X_1(t)}^{X_2(t)} u(t,x)\,dx = \int_{X_1(0)}^{X_2(0)} u(0,x)\,dx .$$

Exercise 8.2. Solve the linear scalar conservation law with flux function $f(x,u) = x^3 u$:

$$\partial_t u + \partial_x(x^3 u) = 0 \ , \qquad u(0,x) = h(x) \ .$$

Is the solution unique? Does $h(x) \in C^\infty$ imply that $u(t,x) \in C^\infty$?

Exercise 8.3. (i) Give the general (entropy condition-satisfying) solution to Burger's equation with the Riemann data

$$\begin{aligned} &u(0,x) = h(x) \ ; \\ &h(x) = h_- \ , \quad x < 0 \ , \qquad h(x) = h_+ \ , \quad x \geq 0 \ . \end{aligned}$$

(ii) Work out the more complicated but still explicit (entropy condition-satisfying) solution to the family of problems with piecewise constant initial data $u(0,x) = h(x)$, where

$$h(x) = h_- \ , \ x < -1 \ , \qquad h(x) = h_0 \ , \ -1 \leq x < 1 \ , \qquad h(x) = h_+ \ , \ x \geq 1 \ .$$

Note: These solutions are global in $(t,x) \in \mathbb{R}^1_+ \times \mathbb{R}^1$, and they involve shock interactions and interactions between shocks and rarefaction waves. There are eight cases depending on the relative inequalities between h_-, h_0, h_+.

Exercise 8.4. On a domain $\Omega \subseteq \mathbb{R}^2$ in which $u(t,x) \in C^1$ solves a general scalar conservation law with convex flux $f(u)$, we showed above that (8.3) is equivalent to Burger's equation, up to transformation of dependent variables. Suppose now that $u(t,x)$ is only a piecewise C^1 weak solution of this conservation law with C^1 shock curves. Under transformation, does it satisfy the jump condition (8.9) for Burger's equation? If not, what condition does it satisfy?

Projects: Chapter 8

Project 8.5 (The Euler equations and shear flows in fluid dynamics). The motion of a fluid is often represented by its velocity field

$$u(t,x) = (u_1(t,x), \ldots, u_d(t,x)) \ ,$$

where $x \in \Omega \subseteq \mathbb{R}^d$, the domain in which the fluid is confined. For obvious physical reasons, one usually considers $d = 2$ or 3. Fluid particles which move according to this velocity field are described by their coordinates $X(t,x)$ which satisfy

$$\frac{dX_j}{dt} = u_j(t, X(t,x)) \ , \qquad X_j(0,x) = x_j \ .$$

The acceleration experienced by this fluid particle is given by

$$\begin{aligned}\frac{d^2 X_j}{dt^2} &= \partial_t u_j(t, X(t,x)) + \sum_{\ell=1}^{d} \partial_{x_\ell} u_j(t, X(t,x)) \frac{dX_j}{dt} \\ &= \big(\partial_t u + (u \cdot \nabla) u\big)_j \ .\end{aligned}$$

An incompressible fluid satisfies

$$\nabla \cdot u = 0 \ ,$$

and it is the most common setting to take the density $\rho = 1$ constant. Newton's laws state that this quantity is proportional to the forces acting on the fluid particle, $\rho(\partial_t u + (u \cdot \nabla)u) = F$. For the simplest incompressible flows with no viscous forces (inviscid flows), an intrinsic component of the force F is given by the pressure

$$F(t, x) = -\nabla p(t, x) \ .$$

When no other forces are acting on the fluid, the equations of motion are called the *Euler equations*

$$\partial_t u + (u \cdot \nabla) u = -\nabla p \ ; \tag{8.26}$$

this is a system of nonlinear equations that involves the components of the d-dimensional vector field $u(t, x)$ and the pressure $p(t, x)$ which is a scalar function. On the boundary $\partial\Omega$ of the fluid domain, the boundary conditions are that $u \cdot N = 0$ where N is the normal to the boundary. The *vorticity* of this fluid vector field is given by

$$\omega(t, x) := \nabla \times u(t, x) \ .$$

(i) Shear flows are special solutions of the Euler equations in the domain $\Omega = \mathbb{R}^d_+$. Considering the case $d = 2$, a shear flow is a velocity vector field given by

$$u(t, x_1, x_2) = (u_1(x_2), 0) \ ,$$

where the function $u_1(x_2)$ is the shear profile. Remark that this flow is time independent, so that if initially $u(t = 0, x)$ is Hölder continuous with exponent $0 \le \alpha \le 1$,

$$|u(t, x) - u(t, x')| \le C|x - x'|^\alpha \quad \forall x, x' \in \mathbb{R}^2_+ \ ,$$

then it remains so for all time $t \in \mathbb{R}^1$.

(a) Show that this vector field is incompressible and that it satisfies the Euler equations (8.26).

(b) Calculate the vorticity of the velocity vector field.

(c) Show that the velocity vector field satisfies the boundary conditions on $\{x_2 = 0\}$.

(ii) Now consider the case of shear flows in three-dimensional settings. In the case that $d = 3$, a shear flow takes the form

$$u(t,x) = (u_1(t,x_2,x_3), u_2(x_3), 0) \ .$$

(a) Show that $\nabla \cdot u = 0$.

(b) Show that if $u(t,x)$ satisfies the Euler equations, then

$$\partial_t u_1 + u_2 \partial_{x_2} u_1 = 0 \ .$$

Therefore $u_1(t,x)$ satisfies a transport equation whose speed is given by $u_2(x_3)$. Show that the solution for $u_1(t,x)$ takes the form

$$u_1(t,x) = u_1(t,x_2,x_3) = u_1(x_2 + tu_2(x_3)) \ .$$

Show that the velocity vector field satisfies the boundary conditions on $\{x_3 = 0\}$. Calculate the vorticity of the flow.

(c) (Hölder modulus of continuity). Suppose that $v(y)$ is a Hölder continuous function of y with Hölder exponent $\beta < 1$,

$$|v(y) - v(y')| \le C_\beta |y - y'|^\beta \ ,$$

and it is composed with a function $u(v)$ which is Hölder continuous with exponent $\gamma < 1$. Then show that the result is in general only Hölder continuous with exponent $\alpha = \beta\gamma < \min\{\beta, \gamma\}$.

(d) (Loss of Hölder continuity by solutions of the Euler equations). This question pursues the fact that in three-dimensional flows, the solution at later time $t \neq 0$ can be less smooth than the initial data for the solution. This phenomenon was originally described in Lions [**15**] and Bardos and Titi [**2**]. Show that if the initial conditions $u(0,x)$ for a three-dimensional shear flow are such that $u_1(0,x_2,x_3)$ is γ-Hölder continuous and $u_2(x_3)$ is β-Hölder continuous, then the solution $u(t,x)$ for $t \neq 0$ need not be any smoother than $\alpha = \beta\gamma < \min\{\beta,\gamma\}$.

Project 8.6 (Moduli of continuity of the solution map of Burger's equation)**.** This project studies several aspects of the dependence of solutions of evolution equations on their initial data. The concept of a well-posed problem is due to Hadamard, who articulated that an initial value problem is *well posed* on a class of functions $\mathcal{C}$ if (i) solutions $u(t)$ exist for data $u_0 \in \mathcal{C}$, (ii) the solution $u(t)$ is uniquely determined by u_0, and (iii) the solution map $\varphi_t : u_0 \mapsto u(t)$ is continuous in the topology given for $\mathcal{C}$.

(i) Consider the solution map for Burger's equation

$$\partial_t u + u\partial_x u = 0 \ , \quad x \in \mathbb{R}^1 \ .$$

Show that the solution map $\varphi_t(u_0)$ is not continuous for u_0 in the space $\mathcal{C}_{piecewise}$ of piecewise continuous functions on $\mathbb{R}^1$. To show this, you might start with a solution that forms a shock, and consider the behavior of nearby initial data and the resulting solutions. If instead the function class one considers is that of continuous functions with piecewise continuous derivatives, is the solution map continuous in this case?

(ii) Show that the solution map $\varphi_t(u_0)$ for Burger's equation is continuous, locally in time, when $\mathcal{C}$ is the Sobolev space $H^s(\mathbb{R}^1)$ for $s > 1$. That is, for $u_0 \in K \subseteq H^s(\mathbb{R}^1)$ a bounded open set, there exists a time $T = T(K) > 0$ such that the initial value problem satisfies the well posedness criterion of Hadamard up to times $|t| < T$.

(iii) However, the solution map on the class $H^s(\mathbb{R}^1)$ is not smooth, nor even Hölder continuous; this is a theorem of Kato [**10**]. His counterexamples to any possible Hölder modulus of continuity are constructed as follows.

(a) Define a family of initial data for Burger's equation to be

$$u_0^\lambda(x) = (\lambda + x_+^{\alpha+1})\varphi(x) ,$$

where $\varphi(x)$ is a C_0^∞ cutoff function such that $\varphi(x) \equiv 1$ for $-1 < x < +1$. The notation is that

$$x_+ = 0 \quad (x < 0) , \quad x_+ = x \quad (x > 0) ,$$

and the exponent $\alpha > 0$ will be taken large, according to the Sobolev regularity exponent s. Show that for $\alpha > s - 1/2$, both $u_0^\lambda(x)$ and $u_0^0(x)$ are $H^s(\mathbb{R}^1)$.

(b) The characteristics for this problem for $\lambda = 0$ and $\lambda > 0$ are, respectively,

$$x = y^0 + u_0^0 t = y^0 + (y_+^0)^{\alpha+1} t , \qquad x = y^0 + u_0^\lambda t = y^\lambda + (\lambda + (y_+^\lambda)^{\alpha+1})t ,$$

where we are restricting our attention to the region where $|x|, |y| < 1$. Show that the respective solutions $u^0(t,x)$ and $u^\lambda(t,x)$ satisfy

$$\begin{aligned} &u^0(t,x) = 0 , \quad x < 0 ; \quad u^0(t,x) = (y_+^0)^{\alpha+1} , \quad 0 \le x ; \\ &u^\lambda(t,x) = \lambda , \quad x < \lambda t ; \quad u^\lambda(t,x) = \lambda + (y_+^\lambda)^{\alpha+1} , \quad \lambda t \le x . \end{aligned}$$

(c) Describe the quantities $y_+^0 = y_+^0(t,x)$ and $y_+^\lambda = y_+^\lambda(t,x)$ in terms of (t,x). In particular, show that

$$x = [1 + (y_+^0)^\alpha t] y^0$$

and therefore for a sufficiently short time interval,

$$(y+^0)^\alpha = \frac{x_+^\alpha}{1 + \alpha x_+^\alpha t} + \mathcal{O}((x_+ t)^2) .$$

Therefore one concludes that $\partial_x^s u^\lambda(t,x) \in L^2(\mathbb{R}^1)$ and the same for $\partial_x^s u^0(t,x)$.

(d) In the space-time region $0 \le x \le \lambda t$, the solution $u^\lambda(t,x)$ is constant while $u^0(t,x)$ is not. Show that

$$\int_0^{\lambda t} |\partial_x^s u^0(t,x)|^2 \, dx \sim (\lambda t)^{2(\alpha - s + 1) + 1} ,$$

and therefore

$$\|u^\lambda(t,\cdot) - u^0(t,\cdot)\|_{H^s}^2 \ge \|\partial_x^s(u^\lambda(t,\cdot) - u^0(t,\cdot))\|_{L^2}^2 \ge C(\lambda t)^{2(\alpha-s+1)+1} .$$

Furthermore, the difference in the initial data is given by

$$\|u_0^\lambda - u_0^0\|_{H^s} = \lambda \|\varphi\|_{H^s} .$$

We may take $(\alpha - s + 1) + \frac{1}{2} = \alpha - s + \frac{3}{2}$ arbitrarily close to zero. The conclusion is that

$$\begin{aligned} \|u^\lambda(t,\cdot) - u^0(t,\cdot)\|_{H^s} &\ge C\lambda^{\alpha - s + \frac{3}{2}} \\ &\ge C\big(\lambda\|\varphi\|_{H^s}\big)^{\alpha - s + \frac{3}{2}} = C\big(\|u_0^\lambda - u_0^0\|_{H^s}\big)^{\alpha - s + \frac{3}{2}} . \end{aligned}$$

Thus the solution map for Burger's equation posed on $H^s(\mathbb{R}^1)$ does not satisfy the criterion for Hölder continuity with exponent $\gamma = \alpha - s + \frac{3}{2}$. Since γ can be made arbitrarily small, such examples exclude the possibility that the solution map is Hölder continuous for any γ. As an extended project, one could ask further whether the solution map satisfies a uniform modulus of continuity that is weaker than Hölder continuity.

The well posedness criterion of Hadamard for evolution equations can be compared with the concept of a *flow* for a dynamical system. Generally for the latter, one asks that a flow satisfy (i) existence of solutions, (ii) uniqueness, and (iii) smoothness of the solution map. The above examples illustrate that a partial differential evolution equation may have a well-posed solution map which may not be a flow by this definition.

Bibliography

[1] B. Abbasi and W. Craig, *On the initial value problem for the wave equation in Friedmann-Robertson-Walker space-times*, Proc. R. Soc. Lond. Ser. A Math. Phys. Eng. Sci. **470** (2014), no. 2169, 20140361, 13, DOI 10.1098/rspa.2014.0361. MR3238168

[2] C. Bardos and E. S. Titi, *Loss of smoothness and energy conserving rough weak solutions for the 3d Euler equations*, Discrete Contin. Dyn. Syst. Ser. S **3** (2010), no. 2, 185–197, DOI 10.3934/dcdss.2010.3.185. MR2610558

[3] R. Beals, *Analysis: An introduction*, Cambridge University Press, Cambridge, 2004. MR2098699

[4] W. Craig and J. Goodman, *Linear dispersive equations of Airy type*, J. Differential Equations **87** (1990), no. 1, 38–61, DOI 10.1016/0022-0396(90)90014-G. MR1070026

[5] H. Dym and H. P. McKean, *Fourier series and integrals*, Probability and Mathematical Statistics, vol. 14, Academic Press, New York-London, 1972. MR0442564

[6] L. C. Evans, *Partial differential equations*, 2nd ed., Graduate Studies in Mathematics, vol. 19, American Mathematical Society, Providence, RI, 2010. MR2597943

[7] G. B. Folland, *Introduction to partial differential equations*, 2nd ed., Princeton University Press, Princeton, NJ, 1995. MR1357411

[8] P. R. Garabedian, *Partial differential equations*, John Wiley & Sons, Inc., New York-London-Sydney, 1964. MR0162045

[9] F. John, *Partial differential equations*, 4th ed., Applied Mathematical Sciences, vol. 1, Springer-Verlag, New York, 1982. MR831655

[10] T. Kato, *The Cauchy problem for quasi-linear symmetric hyperbolic systems*, Arch. Rational Mech. Anal. **58** (1975), no. 3, 181–205, DOI 10.1007/BF00280740. MR0390516

[11] W. Thomson, *On ship waves*, Institution of Mechanical Engineers, Proceedings 38 (1887), 409–434.

[12] S. Klainerman, *Global existence for nonlinear wave equations*, Comm. Pure Appl. Math. **33** (1980), no. 1, 43–101, DOI 10.1002/cpa.3160330104. MR544044

[13] S. Klainerman and P. Sarnak, *Explicit solutions of cmu = 0 on the Friedmann-Robertson-Walker space-times*, Ann. Inst. H. Poincaré Sect. A (N.S.) **35** (1981), no. 4, 253–257 (1982). MR650132

[14] P. Lax, *Hyperbolic Systems of Conservation Laws and the Mathematical Theory of Shock Waves*, CBMS-NSF Regional Conference Series in Applied Mathematics, Society for Industrial and Applied Mathematics (1973). ISBN: 978-0-89871-177-6.

[15] P.-L. Lions, *Mathematical topics in fluid mechanics. Vol. 2. Compressible Models*, Oxford Lecture Series in Mathematics and its Applications, vol. 10, Oxford Science Publications, The Clarendon Press, Oxford University Press, New York, 1998. MR1637634

[16] R. McOwen, *Partial Differential Equations: Methods and Applications, Second Edition*, Pearson (2002). ISBN-10: 0130093351.

[17] L. Nirenberg, *Lectures on linear partial differential equations*, Conference Board of the Mathematical Sciences Regional Conference Series in Mathematics, vol. 17, American Mathematical Society, Providence, R.I., 1973. MR0450755

[18] J. Rauch, *Partial differential equations*, Graduate Texts in Mathematics, vol. 128, Springer-Verlag, New York, 1991. MR1223093

[19] M. E. Taylor, *Pseudodifferential operators*, Princeton Mathematical Series, vol. 34, Princeton University Press, Princeton, N.J., 1981. MR618463

[20] A. N. Tikhonov and V. Y. Arsenin, *Solutions of ill-posed problems*, Scripta Series in Mathematics, V. H. Winston & Sons, Washington, D.C.: John Wiley & Sons, New York-Toronto, Ont.-London, 1977. Translated from the Russian. MR0455365

[21] A. Vasy, *Partial differential equations: An accessible route through theory and applications*, Graduate Studies in Mathematics, vol. 169, American Mathematical Society, Providence, RI, 2015. MR3410751

Index